SpringerBriefs in Information Security and Cryptography

Editor-in-Chief

Yang Xiang, Swinburne University of Technology, Melbourne, Australia

Series Editors

Liqun Chen ⓘ, Department of Computer Science, University of Surrey, Guildford, UK

Kim-Kwang Raymond Choo ⓘ, Department of Information Systems, The University of Texas at San Antonio, San Antonio, USA

Sherman S. M. Chow ⓘ, Chinese University of Hong Kong, Hong Kong, Hong Kong

Robert H. Deng ⓘ, Singapore Management University, Singapore, Singapore

Dieter Gollmann, FB 4-14, TU Hamburg-Harburg, Hamburg, Germany

Kuan-Ching Li, Department of Computer Science and Information Engineering, Providence University, Taichung, Taiwan

Javier Lopez, University of Malaga, Malaga, Spain

Kui Ren, University at Buffalo, Buffalo, USA

Jianying Zhou ⓘ, Singapore University of Technology and Design (SUTD), Singapore, Singapore

The series aims to develop and disseminate an understanding of innovations, paradigms, techniques, and technologies in the contexts of information and cybersecurity systems, as well as developments in cryptography and related studies.

It publishes concise, thorough and cohesive overviews of state-of-the-art topics in these fields, as well as in-depth case studies. The series also provides a single point of coverage of advanced and timely, emerging topics and offers a forum for core concepts that may not have reached a level of maturity to warrant a comprehensive monograph or textbook.

It addresses security, privacy, availability, and dependability issues, also welcoming emerging technologies such as artificial intelligence, cloud computing, cyber physical systems, and big data analytics related to cybersecurity research. Among some core research topics:

Fundamentals and theories

- Cryptography for cybersecurity
- Theories of cybersecurity
- Provable security

Cyber Systems and Secure Networks

- Cyber systems security
- Network security
- Security services
- Social networks security and privacy
- Cyber attacks and defense
- Data-driven cyber security
- Trusted computing and systems

Applications and others

- Hardware and device security
- Cyber application security
- Human and social aspects of cybersecurity

Kwangjo Kim

Practical Post-Quantum Signatures

FALCON and SOLMAE with Python

Kwangjo Kim
School of Computing
KAIST(Korea Advanced Institute for
Science and Technology) /IRCS
(International Research institute for Cyber
Security)
Daejeon, Korea (Republic of)

ISSN 2731-9555　　　　　　　ISSN 2731-9563　(electronic)
SpringerBriefs in Information Security and Cryptography
ISBN 978-3-031-81249-1　　　ISBN 978-3-031-81250-7　(eBook)
https://doi.org/10.1007/978-3-031-81250-7

© The Editor(s) (if applicable) and The Author(s), under exclusive license to Springer Nature Switzerland AG 2025

This work is subject to copyright. All rights are solely and exclusively licensed by the Publisher, whether the whole or part of the material is concerned, specifically the rights of reprinting, reuse of illustrations, recitation, broadcasting, reproduction on microfilms or in any other physical way, and transmission or information storage and retrieval, electronic adaptation, computer software, or by similar or dissimilar methodology now known or hereafter developed.
The use of general descriptive names, registered names, trademarks, service marks, etc. in this publication does not imply, even in the absence of a specific statement, that such names are exempt from the relevant protective laws and regulations and therefore free for general use.
The publisher, the authors and the editors are safe to assume that the advice and information in this book are believed to be true and accurate at the date of publication. Neither the publisher nor the authors or the editors give a warranty, expressed or implied, with respect to the material contained herein or for any errors or omissions that may have been made. The publisher remains neutral with regard to jurisdictional claims in published maps and institutional affiliations.

This Springer imprint is published by the registered company Springer Nature Switzerland AG
The registered company address is: Gewerbestrasse 11, 6330 Cham, Switzerland

If disposing of this product, please recycle the paper.

Preface

The current digital signature methods, such as RSA (Rivest-Shamir-Adleman), DSA (Digital Signature Algorithm), and ECDSA (Elliptic Curve Digital Signature Algorithm), are relatively straightforward in terms of mathematical understanding. While the signing and verification processes differ depending on the key used, the time required for these operations is nearly the same across these algorithms. However, in the era of quantum computing, cryptographic methods must defend against both current classical and future quantum attacks. Achieving this requires a deeper understanding of algebraic geometry, lattice theory, Gaussian sampling, and efficient polynomial computation techniques like FFT (Fast Fourier Transform) and NTT (Number Theoretic Transform), which are crucial for most lattice-based cryptosystems.

The FALCON algorithm, selected as a finalist in the NIST (National Institute of Standards and Technology) Post-Quantum Cryptography (PQC) standardization project after seven years of global evaluation, is a hash-and-sign digital signature scheme based on the NTRU (N-th degree TRUncated polynomial) lattice problem, within the GPV framework. Compared to other quantum-resistant signatures like DILITHIUM and SPHINCS+, FALCON offers a significantly smaller combined size for its public key and signature.

Following FALCON's release, the SOLMAE algorithm was introduced in 2021. Like FALCON, SOLMAE is another hash-and-sign scheme that operates within the GPV (Genry Peikert Vaikuntanathan) framework but simplifies FALCON's complex signing process. Both FALCON and SOLMAE have been implemented in Python, making them easier to understand and work with compared to other low-level programming languages. To enhance understanding of their functionality, we developed Python scripts that examine each algorithm step-by-step, breaking down the results and the underlying mathematical concepts.

This monograph serves as an introductory or educational textbook for undergraduate and graduate students, practitioners, engineers, and anyone interested in post-quantum digital signatures. The material aims to present complex cryptographic concepts in an accessible manner.

This monograph highlights:

- Clarity and Focus: The title clearly indicates that the book is about *post-quantum signatures*, which is a critical and timely topic in the field of modern cryptography.
- Practical Approach: The word *Practical* suggests that the book will focus on hands-on, applicable knowledge, which is appealing to readers looking to implement these cryptographic techniques.
- Specific Algorithms: Mentioning FALCON and SOLMAE directly in the title highlights the specific post-quantum signature schemes covered, attracting readers who are specifically interested in these algorithms.
- Programming Language: Including *with Python* informs readers that the book will provide implementation examples or exercises using Python, which is a popular language for such purposes.

YongIn, Republic of Korea Kwangjo Kim
December 2024

Acknowledgments

The author expresses his sincere gratitude to the volunteer group who designed the initial version of SOLMAE. This group includes Mehdi Tibouchi, Alexandre Wallet, Thomas Espitau, Akira Takahashi, and Yang Yu. The author is also grateful to Seungki Kim and YeonJun Kim for their contributions to preparing the revised specification of SOLMAE.

I would like to express my gratitude to my high school friends who accompanied me and helped inspire the concept for this monograph in part while trekking Mont Blanc for a week, starting in Chamonix, France, in August 2014.

I dedicate this monograph to my late parents, whose love and sacrifices laid the foundation for my strength and resilience. Thanks to my wife, Nami Jang and my family for their strong support. On the hill overlooking Seoksung Mt. in Yongin.

Contents

1 **Introduction** 1
2 **Notations and Definition** 5
3 **FALCON Algorithm** 11
4 **SOLMAE Algorithm** 19
5 **Basics of Python** 27
6 **Checking FALCON with Python** 31
7 **Checking SOLMAE with Python** 61
8 **Concluding Remarks** 79

References 81
Index 85

About the Author

Kwangjo Kim received his B.Sc. and M.Sc. degrees in Electronic Engineering from Yonsei University, Korea, in 1980 and 1983, respectively, and his Ph.D. from the Division of Electrical and Computer Engineering, Yokohama National University, Japan, in 1991. From 1979 to 1997, he worked at the Electronics and Telecommunications Research Institute (ETRI), serving as the Head of Coding Section I. He has held various prestigious visiting positions, including Visiting Professor roles at the Massachusetts Institute of Technology (MIT), the University of California at San Diego (UCSD) in 2005, the Khalifa University of Science, Technology, and Research (KUSTAR), UAE, in 2012, and the Bandung Institute of Technology (ITB), Indonesia, in 2013. After retiring in August 2021 from the Korea Advanced Institute of Science and Technology (KAIST), Korea, where he had worked since 1998, Professor Kim has been serving as the President of the International Research Institute for Cyber Security (IRCS, https://ircs.re.kr), a non-profit organization approved by the Korean government. He is also an Emeritus Professor at the School of Computing and the Graduate School of Information Security at KAIST and the Honorary President of the Korea Institute of Information Security and Cryptography (KIISC). Professor Kim has made significant contributions to the field of cryptography. He served as a Board Member of the International Association for Cryptologic Research (IACR) from 2000 to 2004 and as Chairperson of the Asiacrypt Steering Committee from 2005 to 2008. He served as the President of KIISC in 2009 and the Korean representative to IFIP TC-11 from 2017 to 2021. Recently he was appointed as Adjunct Faculty in the Department of Electrical and Computer Engineering at Cleveland State University for the fiscal years 2025 through 2028.

He was honored as the first Korean Fellow of the IACR for his contributions to cryptographic design, education, and leadership, and for his exemplary service to the IACR and the Asia-Pacific cryptographic community. In addition to his leadership roles, Professor Kim served as General Chair for Asiacrypt 2020 (online) and PQCrypto 2021 (hybrid), both held in Daejeon, Korea, including CHES2014 in Busan, Korea, and Asiacrypt2004, Jeju Island, Korea, etc.

He coauthored with M.E. Aminanto, and H.C.Tanuwidjaja, *Network Intrusion Detection using Deep Learning – A Feature Learning Approach* in 2018 and with H.C. Tanuwidjaja *Privacy-Preserving Deep Learning – A Comprehensive Survey* in 2021, both published by Springer Briefs on Cyber Security Systems and Networks.

He was recognized as one of the World's Top 2% Scientists by Stanford University in 2023 and a key figure in the implementation of SOLMAE in Python, a quantum-secure signature scheme that is faster and more efficient than FALCON, which was selected as a FIPS standard by NIST in 2021. Professor Kim has an H-index of 48 with 10,295 citations according to Google Scholar in 2024, and his most cited paper is "ID-based blind signature and ring signature from pairings," coauthored with F. Zhang and presented at Asiacrypt 2002, which has been cited 814 times.

Professor Kim's current research interests include cryptologic theory and practice, cybersecurity, and their applications, holding 10 international patents and 20 domestic patents.

For more details, please visit: https://caislab.kaist.ac.kr/kkj.

Acronyms

CRYSTALS	CRYptographic SuiTe for Algebraic Lattices
DH	Diffie Hellmam
DS	Digital Signature
DSA	Digital Signature Algorithm
ECDSA	Elliptic Curve Digital Signature Algorithm
FALCON	FAst Fourier Lattice-based COmpact signatures over NTRU
FFT	Fast Fourier Transform
FIPS	Federal Information Processing Standard
GPV	Gentry Peikert Vaikuntanathan
GSO	Gram Schmidt Orthogonalization
KAT	Known Answer Test
KEM	Key Encapsulation Mechanism
NIST	National Institute of Standards and Technology
NTRU	Number Theory aRe Us or N-th degree TRUncated polynomial
NTT	Number Theoretic Transform
PQC	Post-Quantum Cryptography
RSA	Rivest Shamir Adelmann
SAGA	Statistically Acceptable GAussians
SOLMAE	Secure algOrithm for Long-term Message Authentication and Encryption

List of Algorithms

1 KeyGen of FALCON 13
2 Sign of FALCON 15
3 Compress 16
4 Decompress 16
5 Verif of FALCON 17
6 KeyGen of SOLMAE 22
7 Sign of SOLMAE 24
8 Verif of SOLMAE 25

List of Figures

Fig. 3.1	Genealogic tree of FALCON	12
Fig. 3.2	Flowchart of `KeyGen` for FALCON	13
Fig. 3.3	Flowchart of `Sign` for FALCON	15
Fig. 4.1	Overview of SOLMAE	20
Fig. 4.2	Flowchart of `KeyGen` for SOLMAE	21
Fig. 4.3	Flowchart of `Sign` for SOLMAE	23
Fig. 5.1	Screen capture of Visual Studio Code	29
Fig. 5.2	Installed packages in my PC environment	30
Fig. 6.1	Output of `test_split_and_merge()`	34
Fig. 6.2	`phi16_roots` used in Script 6.3 for FFT	35
Fig. 6.3	Output of checking `ftt.py`	37
Fig. 6.4	`phi16_roots` used in Script 6.6 for NTT	39
Fig. 6.5	Output of checking `ntt.py`	39
Fig. 6.6	Output of checking `ntrugen.py`	40
Fig. 6.7	Output of six test cases	42
Fig. 6.8	Description of parameters	43
Fig. 6.9	Specific values of various parameters for both FALCON-512 and FALCON-1024	44
Fig. 6.10	Output of `test_samplerz`	45
Fig. 6.11	Comparison of generated random integers with ideal Gaussian	46
Fig. 6.12	Output of 5 `ffnp()` tests for FALCON-512	48
Fig. 6.13	Output of 5 `ffnp()` tests for FALCON-1024	49
Fig. 6.14	Three examples of key pairs and signature executing FALCON-512	51
Fig. 6.15	Three examples of key pairs and signature executing FALCON-1024	52
Fig. 6.16	Specification of my test computer used in `test.py`	52
Fig. 6.17	Time consumed in msec executing `test.py`	59
Fig. 7.1	Output of `solmae_params.py`	65
Fig. 7.2	Scatter plot of `Unifcrown.py`	66

Fig. 7.3	Scatter and QQ plots of checking `N_sampler.py`	68
Fig. 7.4	Sample output from executing `Pairgen.py` for SOLMAE-512	70
Fig. 7.5	Sample output from executing `Pairgen.py` for SOLMAE-1024	71
Fig. 7.6	Sample output by executing `keygen.py` for SOLMAE-512	74
Fig. 7.7	Sample output by executing `keygen.py` for SOLMAE-1024	74
Fig. 7.8	Two tests of keygen, sign and verify procedures of SOLMAE-512	77
Fig. 7.9	Two tests of keygen, sign and verify procedures of SOLMAE-1024	78

Chapter 1
Introduction

The history of cryptography began with simple substitution of plaintext letters, such as the Caesar cipher [20]. During World War I and World War II, mechanical encryption devices using such as multiple rotors were employed. The modern cryptography has started from 1949 influenced by Shannon's seminal paper entitled as "Communication Theory of Secrecy Systems" [42].

The goal of modern cryptography is to secure information transmission, storage, and processing through the Internet or other channels, protecting against illegal eavesdropping, tampering, and forgery by unauthorized and malicious third parties. This cryptography involves an encryption process that takes plaintext and a key as inputs, and a decryption process that takes the ciphertext and a key as inputs to produce a decrypted message identical to the original plaintext. If the same key is used for both processes, it is called secret key or symmetric key cryptography. If different keys are used, one of which is made public, the other kept in private (or secret), it is called public key or asymmetric key cryptography.

Cryptography provides several security services, such as Confidentiality, which ensures secure communication over a transmission medium or channel even if a third party is eavesdropping. It also provides Authentication, which protects against intentional tampering or forgery of messages by third parties, guaranteeing message Integrity and including Identification, which distinguishes between legitimate and illegitimate entities. Cryptology consists of both cryptography, the design of cryptographic systems that are secure against various known attack methods, and cryptanalysis, which seeks to discover secret information by using attack algorithms and tools against publicly available data. Cryptographic designers must ensure that their systems are secure not only against known attacks but also against future attacks during the expected usage period. Cryptanalysts, on the other hand, use the best available computers to exploit cryptographic vulnerabilities and statistical properties to recover plaintext or secret keys from ciphertext within a feasible timeframe. For example, an n-bit secret key encryption requires a search space of $O(2^n)$, and if this search space cannot realistically be explored with current

© The Author(s), under exclusive license to Springer Nature Switzerland AG 2025
K. Kim, *Practical Post-Quantum Signatures*, SpringerBriefs in Information Security and Cryptography, https://doi.org/10.1007/978-3-031-81250-7_1

technology, the cryptographic system is considered to be secure. Currently, the minimum key size is 128 bits, and up to 256 bits is commonly used. NIST has standardized AES-128 and AES-256 algorithms in FIPS 197 [31], which are widely used globally.

On the other hand, public key cryptography was first devised in 1976 by Diffie and Hellman(DH) [1] as a method where two parties who wish to communicate publicly by exchanging public keys each other, and using their private keys, they generate a shared secret known only to them.

The security of this DH method relies on the fact that it must be infeasible to derive the private key from the public key, a problem known as the discrete logarithm problem. To remain secure against digital computer attacks, the modulus used in DH must be at least 2048 bits and can go up to 8192 bits.

In 1977, RSA (Rivest Shamir Adleman) [40] method extended the DH method by proposing the RSA public key cryptosystem, which uses the product of two large prime numbers for modular exponentiation in the encryption and decryption processes. RSA's security is based on the difficulty of the prime factorization problem, and currently, modulus sizes range from 2048 bits to 8192 bits to ensure security against digital attacks. In 1985 and 1987, Miller [29] and Koblitz [25] simultaneously proposed a key exchange system using elliptic curves, which reduces the modulus size of the DH method by a factor of six. This method is based on the elliptic curve discrete logarithm problem, and key sizes currently range from 256 bits to 521 bits.

Public key cryptography systems like RSA method are categorized into two types based on key usage: Key Encapsulation Mechanisms (KEM), used to securely share a randomly generated session key, and Digital Signatures (DS), which ensure message integrity. In KEM, a random session key is encrypted using the recipient's public key and decrypted using their private key to share the session key. In DS, the sender (or signer) hashes the message to create a digest and then signs it using their private key. The recipient (or verifier) verifies the signature using the public key. If the verification matches, the message is accepted; otherwise, it is rejected. Thus, the sender signs the message with a private key, while the recipient verifies the signature with a public key. The current digital signature (DS) methods are categorized into the hash-and-sign [14] method and the Fiat-Shamir-with-aborts [27] method.

The RSA-based signature method poses risks of forgery, so ElGamal [5] proposed a probabilistic signature scheme based on the discrete logarithm problem. NIST adopted this as a standard algorithm and defined the Digital Signature Algorithm (DSA) in FIPS 186-4 [32], recommending a key size of at least 2048 bits. NIST also extended DSA to elliptic curves in FIPS 186-4 [32], establishing the ECDSA algorithm, with key sizes ranging from 256 bits to 521 bits.

In 1999, Shor [43] proposed an efficient randomized algorithm on a hypothetical quantum computer in 1999 to integer factorization and discrete logarithm problems in a polynomial time. Building for the powerful computing environment at that time was beyond imagination. Currently the threat of attacking the current (or classical) secure system by using the quantum computer is expected to be right at our fingertips due to the aggressive road map by IBM quantum computing [19].

We are very concerned about so-called *Harvest Now, Decrypt Later* attack [45] which is a surveillance strategy that relies on the acquisition and long-term storage of currently unreadable encrypted data awaiting possible breakthroughs in decryption technology that would render it readable in the future.

Due to the substantial amount of research on quantum computers, large-scale quantum computers if built, can break many public-key cryptosystems based on the number-theoretic hard problems in use. In 2016, NIST [37] has initiated Post Quantum Cryptography (PQC) project to solicit, evaluate, and standardize one or more quantum-resistant cryptographic algorithms for KEM and DS globally. After several rounds, NIST has finally selected CRYSTALS-KYBER [41] for KEM and CRYSTALS-DILITHIUM [28], FALCON [9][1] and SPHINCS+ [18] for DS in 2022. The FIPS PUB standard of KYBER, DILITHIUM and SPHINCS+ are available at [33, 34] and [35], respectively in 2024. As of writing this monograph, the FIPS PUB standard of FALCON is under process.

Influenced by this NIST PQC project, Korean cryptographic society led by KpqC task force [26] has called for soliciting Korean PQC candidates by the end of Oct. in 2022. By the due of submission, 7 candidates KEM and 8 candidates DS for KpqC competition were submitted and their details are available at https://kpqc.or.kr/.

SOLMAE which stands for an acronym of quantum-**S**ecure alg**O**rithm for **L**ong-term **M**essage **A**uthentication and **E**ncryption was submitted to KpqC Competition as one of DS candidate algorithms which is a lattice-based signature scheme inspired by several pioneering works based on the hash-then-sign signature paradigm proposed by Gentry et al. [12]. SOLMAE is inspired from FALCON's design. Some of the new theoretical foundations were laid out in the presentation of MITAKA [7] while keeping the security level of FALCON with 5 NIST levels of security I to V. At a high level, SOLMAE removes the inherent technicality of the sampling procedure, and most of its induced complexity from an implementation standpoint, for *free*, that is with no loss of efficiency. This theoretical simplicity translates into faster operations while preserving signatures and verification key sizes, on top of allowing for additional features absent from FALCON, such as enjoying cheaper masking and being parallelizable.

RSA and DH methods can be understood through number theoretic knowledge, but to comprehend FALCON and SOLMAE in depth, not only is it necessary to understand algebraic knowledge and Gaussian sampling techniques, but also a foundational understanding of lattice theory and polynomial arithmetics. Even after reading the specifications for both signing methods, a significant amount of mathematical background knowledge is required, making it difficult to grasp. Since Python packages that implement both methods are publicly available, this monograph minimizes the mathematical explanations and instead aims to help anyone understand these two methods through Python scripts easily.

The organization of this monograph is as follows: In Chap. 2, we define our notations and definition used in this monograph. In Chaps. 3 and 4, we overview

---

[1] It stands for the acronym: **Fa**st **F**ourier **l**attice-based **co**mpact signatures over **NTRU**.

the specification of FALCON and SOLMAE including their keygen, signing and verification procedures, respectively. After introducing the basics of Python and how to set up your test environment over Windows OS in Chap. 5, we describe how to perform the correctness of the common modules and functions, `common.py`, `fft.py`, `ntt.py`, `ntrugen.py` and `encoding.py` both used in FALCON and SOLMAE and check their correctness used only in FALCON, `parameters.py`, `samplerz.py`, `ffnp()` in `ffsampling.py`, `falcon.py` and `test.py` in Chap. 6. In Chap. 7, we describe the correctness of specific functions used in SOLMAE only which include `parameters.py`, `Unifcrwon.py`, `N_sampler.py`, `Parigen.py`, `keygen.py` and `solmae.py` by omitting the description of common modules used for FALCON and SOLMAE.

Finally, we will give concluding remarks and suggest challenging issues in Chap. 8.

Chapter 2
Notations and Definition

2.1 Matrices, Vectors, and Scalars

Matrices will usually be in bold uppercase (e.g., **B**), vectors in bold lowercase (e.g., **v**), and scalars—which include polynomials—in italic (e.g. s). We use the row convention for vectors. The transpose of a matrix **B** may be noted $\mathbf{B}^t$. The ℓ_2-norm of a vector $\mathbf{x} = (x_1, \ldots, x_d)$ is $\|\mathbf{x}\| = \left(\sum_i |x_i|^2\right)^{1/2}$ and its ℓ_∞-norm is $\|\mathbf{x}\|_\infty = \max_i |x_i|$. It is to be noted that for a polynomial f, we do *not* use f' to denote its derivative in this monograph.

2.2 Quotient Ring

Let $\mathbb{Z}$ and $\mathbb{N}$ denote a set of integers and a set of all numbers starting from 1, respectively. $\mathbb{Q}$ and $\mathbb{R}$ denote a set of rational numbers and a set of real numbers, respectively. For $q \in \mathbb{N}^\times$, we denote by $\mathbb{Z}_q$ the quotient ring $\mathbb{Z}/q\mathbb{Z}$. In FALCON and SOLMAE, an integer modulus $q = 12,289$ is prime, so $\mathbb{Z}_q$ is also a finite field. We denote by $\mathbb{Z}_q^\times$ the group of invertible elements of $\mathbb{Z}_q$, and by φ Euler's totient function: $\varphi(q) = |\mathbb{Z}_q^\times| = q - 1 = 3 \cdot 2^{12}$ since q is prime. The rings $\mathbb{Q}[x]/(\phi)$, $\mathbb{Z}[x]/(\phi)$, and $\mathbb{R}[x]/(\phi)$ where ϕ is a monic minimal polynomial will be interchangeably written as $\mathcal{Q}$, $\mathcal{Z}$, and $K_\mathbb{R}$, respectively for the sake of our convenience.

2.3 Number Fields

Let $a = \sum_{i=0}^{d-1} a_i x^i$ and $b = \sum_{i=0}^{d-1} b_i x^i$ be arbitrary elements of the number field $\mathcal{Q} = \mathbb{Q}[x]/(\phi)$. We note a^* and call (Hermitian) adjoint of a the unique element of Q such that for any root ζ of ϕ, $a^*(\zeta) = \overline{a(\zeta)}$, where $\overline{\cdot}$ is the usual complex conjugation over $\mathbb{C}$. For $\phi = x^d + 1$, the Hermitian adjoint a^* can be expressed simply:

$$a^* = a_0 - \sum_{i=1}^{d-1} a_i x^{d-i} \qquad (2.1)$$

We extend this definition to vectors and matrices: the adjoint $\mathbf{B}^*$ of a matrix $\mathbf{B} \in Q^{n \times m}$ (resp. a vector $\mathbf{v}$) is the component-wise adjoint of the transpose of $\mathbf{B}$ (resp. $\mathbf{v}$):

$$\mathbf{B} = \begin{bmatrix} a & b \\ c & d \end{bmatrix} \Leftrightarrow \mathbf{B}^* = \begin{bmatrix} a^* & c^* \\ b^* & d^* \end{bmatrix} \qquad (2.2)$$

2.4 Inner Product

The inner product $\langle \cdot, \cdot \rangle$ over Q and its associated norm $\|\cdot\|$ are defined as:

$$\langle a, b \rangle = \frac{1}{\deg(\phi)} \sum_{0 < i \leq d} \varphi_i(a) \cdot \overline{\varphi_i(b)} \qquad (2.3)$$

$$\|a\| = \sqrt{\langle a, a \rangle} \qquad (2.4)$$

These definitions can be extended to vectors: for $u = (u_i)$ and $v = (v_i)$ in Q^m, $\langle u, v \rangle = \sum_i \langle u_i, v_i \rangle$. For our choice of ϕ, the inner product coincides with the usual coefficient-wise inner product:

$$\langle a, b \rangle = \sum_{0 \leq i < d} a_i b_i; \qquad (2.5)$$

From an algorithmic point of view, computing the inner product or the norm is most easily done using Eq. (2.3) if polynomials are in FFT representation, and using Eq. (2.5) if they are in coefficient representation. By substituting $b = a$ in Eqs. (2.3) and (2.5), we get

$$\|\varphi(a)\| = \sqrt{d} \cdot \|a\|. \qquad (2.6)$$

where $\|\cdot\|$ is Euclidean norm. Since we know that

$$\|\varphi(a)\| = \sqrt{2} \cdot \|(Re(\varphi_1(a)), Im(\varphi_1(a)), \cdots Re(\varphi_{d/2}(a)), Im(\varphi_{d/2}(a)))\|, \qquad (2.7)$$

we get

$$\|(Re(\varphi_1(a)), Im(\varphi_1(a)), \cdots Re(\varphi_{d/2}(a)), Im(\varphi_{d/2}(a)))\| = \sqrt{\frac{d}{2}} \cdot \|a\|. \quad (2.8)$$

If $a \in K_{\mathbb{R}}$ follows the d-dimensional standard normal distribution, it is known that

$$(Re(\varphi_1(a)), Im(\varphi_1(a)), \cdots Re(\varphi_{d/2}(a)), Im(\varphi_{d/2}(a))) \text{ follows } \mathcal{N}_{d/2}, \quad (2.9)$$

where $\mathcal{N}_{d/2}$ denotes continuous Gaussian distribution with zero mean and $\frac{d}{2} \cdot I_d$ (i.e., Identity matrix) variance.

2.5 Lattice

A lattice is a discrete subgroup of $\mathbb{R}^n$. Equivalently, it is the set of *integer* linear combinations obtained from a basis **B** of $\mathbb{R}^n$. The volume of a lattice is det **B** for any of its basis.

2.6 Ring Lattices

For the rings $Q = \mathbb{Q}[x]/(\phi)$ and $\mathcal{Z} = \mathbb{Z}[x]/(\phi)$, positive integers $m \geq n$, and a full-rank matrix $\mathbf{B} \in Q^{n \times m}$, we denote by $\Lambda(\mathbf{B})$ and call lattice generated by **B**, the set $\mathcal{Z}^n \cdot \mathbf{B} = \{z\mathbf{B} \mid z \in \mathcal{Z}^n\}$. By extension, a set Λ is a lattice if there exists a matrix **B** such that $\Lambda = \Lambda(\mathbf{B})$. We may say that $\Lambda \subseteq \mathcal{Z}^m$ is a q-ary lattice if $q\mathcal{Z}^m \subseteq \Lambda$.

2.7 NTRU Lattices

Let q be an integer, and $f \in \mathbb{Z}[x]/(x^d + 1)$ such that f is invertible modulo q (equivalently, $\det[f]$ is coprime to q). Let $h = g/f \mod q$ and consider the NTRU module associated to h:

$$\mathcal{M}_{\text{NTRU}} = \{(u, v) \in K_{\mathbb{R}}^2 : hu - v = 0 \mod q\},$$

and its lattice version

$$\mathcal{L}_{\text{NTRU}} = \{(\mathbf{u}, \mathbf{v}) \in \mathbb{Z}^{2d} : [h]\mathbf{u} - \mathbf{v} = 0 \mod q\}.$$

This lattice has volume q^d. Over $K_\mathbb{R}$, it is generated by (f, g) and any (F, G) such that $fG - gF = q$. For such a pair $(f, g), (F, G)$, this means that $\mathcal{L}_{\text{NTRU}}$ has a basis of the form

$$\mathbf{B}_{f,g} = \begin{bmatrix} [f] & [F] \\ [g] & [G] \end{bmatrix}.$$

One checks that $([h], -\text{Id}_d) \cdot \mathbf{B}_{f,g} = 0 \mod q$, so the verification key is h. The NTRU-search problem is: given $h = g/f \mod q$, find any $(f' = x^i f, g' = x^i g)$. In its decision variant, one must distinguish $h = g/f \mod q$ from a uniformly random $h \in R_q := \mathbb{Z}[x]/(q, x^d + 1) = (\mathbb{Z}/q\mathbb{Z})[x]/(x^d + 1)$. These problems are assumed to be intractable for large d.

2.8 DFT Representation

For $d = 2^n$, we use $\phi(x) = x^d + 1$. It is a monic polynomial of $\mathbb{Z}[x]$, irreducible in $\mathbb{Q}[x]$ and with distinct roots over $\mathbb{C}$. Then $\zeta_j = exp(i(2j - 1)\pi/d)$ for $j = 1, 2, \cdots d$ are roots of $\phi(x)$. For $f = \Sigma f_i x^i \in K_\mathbb{R}$, we define the coefficient representation as $\mathbf{f} = (f_0, f_1, \cdots f_{d-1})$ and Discrete Fourier Transform (DFT) representation $\varphi(f) = (\varphi_1(f), \cdots, \varphi_d(f))$.

2.9 Discrete Gaussians

For $\sigma, \mu \in \mathbb{R}$ with $\sigma > 0$, we define the Gaussian function $\rho_{\sigma,\mu}$ as $\rho_{\sigma,\mu}(x) = exp(-|x-\mu|^2/2\sigma^2)$, and the discrete Gaussian distribution $D_{\mathbb{Z},\sigma,\mu}$ over the integers as:

$$D_{\mathbb{Z},\sigma,\mu}(x) = \frac{\rho_{\sigma,\mu}(x)}{\sum_{z \in \mathbb{Z}} \rho_{\sigma,\mu}(z)} \quad (2.10)$$

The parameter μ may be omitted when it is equal to zero.

2.10 Gram-Schmidt Orthogonalization

Any matrix $\mathbf{B} \in Q^{n \times m}$ can be decomposed as follows:

$$\mathbf{B} = \mathbf{L} \times \tilde{\mathbf{B}} \quad (2.11)$$

where **L** is lower triangular with 1's on the diagonal, and the rows $\tilde{b}_i$'s of $\tilde{\mathbf{B}}$ verify $\langle \tilde{b}_i, \tilde{b}_j \rangle = 0$ for $i \neq j$. When **B** is full-rank, this decomposition is unique, and it is called the Gram-Schmidt orthogonalization (or GSO). We also call the Gram-Schmidt norm of **B** the following value:

$$\|\mathbf{B}\|_{GS} = \max_{\tilde{\mathbf{b}}_i \in \tilde{\mathbf{B}}} \|\tilde{\mathbf{b}}_i\| \qquad (2.12)$$

2.11 LDL* Decomposition

The LDL* decomposition writes any full-rank Gram matrix as a product LDL*, where $\mathbf{L} \in Q^{n \times n}$ is lower triangular with 1's on the diagonal, and $\mathbf{D} \in Q^{n \times n}$ is diagonal. The LDL* decomposition and the GSO are closely related as for a basis **B**, there exists a unique GSO $\mathbf{B} = \mathbf{L} \cdot \tilde{\mathbf{B}}$, and for a full-rank Gram matrix **G**, there exists a unique LDL* decomposition $\mathbf{G} = \mathbf{LDL}^*$. If $\mathbf{G} = \mathbf{BB}^*$, then $\mathbf{G} = \mathbf{L} \cdot (\tilde{\mathbf{B}}\tilde{\mathbf{B}}^*) \cdot \mathbf{L}^*$ is a valid LDL* decomposition of **G**. As both decompositions are unique, the matrices **L** in both cases are actually the same. In a nutshell:

$$[\mathbf{L} \cdot \tilde{\mathbf{B}} \text{ is the GSO of } \mathbf{B}] \Leftrightarrow [\mathbf{L} \cdot (\mathbf{B}\tilde{\mathbf{B}}^*) \cdot \mathbf{L}^* \text{ is the LDL}^* \text{ decomposition of } (\mathbf{BB}^*)]. \qquad (2.13)$$

The reason why we present both equivalent decompositions is that the GSO is a more familiar concept in lattice-based cryptography, whereas the use of LDL* decomposition is faster and therefore makes more sense from an algorithmic point of view.

Chapter 3
FALCON Algorithm

3.1 Overview

Hoffstein et al. [16] suggested a new public-key cryptosystem based on a polynomial ring in 1997 as an alternative to RSA and DH

whose difficulties are based on number-theoretic hard problems such as integer factorization and discrete log problem, respectively. They founded the company so-called as NTRU[1] Cryptosystem with Lieman and initiated an open-source lattice-based cryptography consisting of two algorithms: NTRUENCRYPT used for encryption/decryption and NTRUSIGN used for digital signatures. Their security relies on the presumed difficulty of factoring certain polynomials in a truncated polynomial ring into a quotient of two polynomials having very small coefficients.

NTRUSIGN was designed based on the GGH signature scheme [13] which was proposed in 1995 based on solving the Closest Vector Problem (CVP) in a lattice and asymptotically is more efficient than RSA in the computation time for encryption, decryption, signing, and verifying are all quadratic in the natural security parameter. The signer demonstrates knowledge of a good basis for the lattice by using it to solve CVP on a point representing the message; the verifier uses a bad basis for the same lattice to verify that the signature under consideration is actually a lattice point and is sufficiently close to the message point.

On the other hand, Min et al. [30] suggested a weak property of malleability of NTRUSIGN using the annihilating polynomial from a given message and signature pair to generate a valid signature. Nguyen and Regev [36] had cryptanalyzed the original GGH signature scheme including NTRUSIGN in 2006 successfully extracting secret information from many known signatures characterized by mul-

---

[1] Number Theorists 'R' Us, or Number Theory Research Unit, or N-th degree TRuncated polynomial Ring.

tivariate optimization problems. Their experiments showed that 90,000 signatures are sufficient to recover the NTRUSIGN-251 secret key.

In a nutshell, FALCON follows a framework introduced in 2008 by Gentry et al. [12] which we call the GPV framework for short over the NTRU lattices and uses a typically hash-and-sign paradigm. Their high-level idea is the following:

1. The public key is a long basis of a q-ary lattice.
2. The private key is (essentially) a short basis of the same lattice.
3. In the signing procedure, the signer:
   (a) generates a random value, $salt$;
   (b) computes a target $\mathbf{c} = H(M\|salt)$, where H is a hash function sending input to a random-looking point (on the grid);
   (c) uses his knowledge of a short basis to compute a lattice point $\mathbf{v}$ close to the target $\mathbf{c}$;
   (d) outputs $(salt, \mathbf{s})$, where $\mathbf{s} = \mathbf{c} - \mathbf{v}$.
4. The verifier accepts the signature $(salt, \mathbf{s})$ if and only if:
   (a) the vector $\mathbf{s}$ is short;
   (b) $H(M\|salt) - \mathbf{s}$ is a point on the lattice generated by his public key.

Only the signer should be able to *efficiently* compute v close enough to an arbitrary target. This is a decoding problem that can be solved when a basis of *short* vectors is known. On the other hand, anyone wanting to check the validity of a signature should be able to verify lattice membership. The **KeyGen**, **Sign** and **Verif** procedures for FALCON will be introduced briefly in the later Section by restating the original specification as in [9]. Fig. 3.1 shows the genealogic tree of FALCON. For details, the readers can refer to [9].

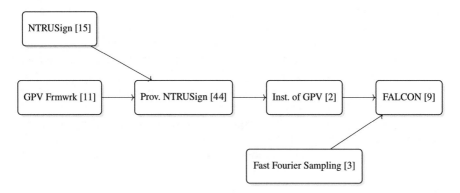

Fig. 3.1 Genealogic tree of FALCON

3.2 Key Generation of FALCON

For the class of NTRU lattices, a trapdoor pair is $(h, \mathbf{B}_{f,g})$ where $h = f^{-1}g$, $\mathbf{B}_{f,g}$ is a trapdoor basis over $\mathcal{L}_{\text{NTRU}}$ and Pornin and Prest [38] showed that a completion (F, G) can be computed in $O(d \log d)$ time from short polynomials $f, g \in \mathcal{Z}$. In practice, their implementation is as efficient as can be for this technical procedure: it is called NtruSolve in FALCON. Their algorithm only depends on the underlying ring and has now a stable version for $\mathbb{Z}[x]/(x^d + 1)$, where $d = 2^n$.

Fig. 3.2 illustrates the flowchart of the key generation procedure for FALCON. Algorithm 1 describes the pseudo-code for key generation of FALCON. Readers can refer to Algorithms 5 and 6 in [9] for details on how to perform ntrugen and ntrusolve, respectively. Additionally, Algorithms 8 and 9 in [9] explain the procedures for LDL* and ffLDL*, respectively.

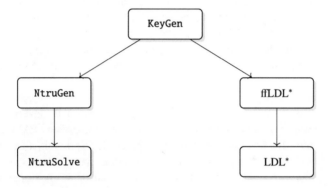

Fig. 3.2 Flowchart of KeyGen for FALCON

Algorithm 1 KeyGen of FALCON

Input: A monic polynomial $\phi \in \mathbb{Z}[x]$, a modulus q
Output: A secret key sk, a public key pk
1: $f, g, F, G \leftarrow \text{NtruGen}(\phi, q)$ // Solving the NTRU equation
2: $\mathbf{B} \leftarrow \begin{bmatrix} g & -f \\ G & -F \end{bmatrix}$;
3: $\hat{\mathbf{B}} \leftarrow \text{FFT}(\mathbf{B})$ // Compute FFT for each $\{g, -f, G, -F\}$
4: $\mathbf{G} \leftarrow \hat{\mathbf{B}} \times \hat{\mathbf{B}}^*$;
5: $T \leftarrow \text{ffLDL}^*(\mathbf{G})$ // Compute the LDL* tree
6: for each leaf of T do
7:      $leaf.value \leftarrow \sigma/\sqrt{leaf.value}$ // Normalization step
8: sk $\leftarrow (\hat{\mathbf{B}}, T)$;
9: $h \leftarrow gf^{-1} \mod q$;
10: pk $\leftarrow h$;
11: **return** sk, pk

3.3 Signing of FALCON

At a high level, the signing procedure in FALCON is at first to compute a hashed value $c \in \mathbb{Z}_q[x]/(\phi)$ from the message, M and a salt r, then using the secret key, f, g, F, G to generate two short values (s_1, s_2) such that $s_1 + s_2 h = c \mod q$. An interesting feature is that only the *first half* of the signature (s_1, s_2) needs to be sent along the message, as long as h is available to the verifier. This comes from the identity $h s_1 = s_2 \mod q$ defining these lattices, as we will see in the Verif algorithm description.

The core of FALCON signing is to use ffSampling (Algorithm 11 in [9]) which applies a randomizing rounding according to Gaussian distribution on the coefficient of $t = (t_0, t_1) \in (\mathbb{Q}[x]/(\phi))^2$ stored in the FALCON Tree, T at the KeyGen procedure of FALCON.

This fast Fourier sampling algorithm can be seen as a recursive version of Klein's well-known trapdoor sampler, but *cannot be computed in parallel* also known as the GPV sampler. Klein's sampler uses a matrix L and the norm of Gram-Schmidt vectors as a trapdoor while FALCON are using a tree of non-trivial elements in such matrices. Note that Fouque et al. [10] suggested Gram-Schmidt norm leakage in FALCON by timing side channels in the implementation of the one-dimensional Gaussian samplers.

FALCON cannot output two different signatures for a message. This well-known concern of the GPV framework can be addressed in several ways, for example, making a stateful scheme or by hash randomization. FALCON chose the latter solution for efficiency purposes. In practice, Sign adds a random "salt" $r \in \{0, 1\}^k$, where k is large enough that an unfortunate collision of messages is unlikely to happen, that is, it hashes $(r||M)$ instead of M. A signature is then $\text{sig} = (r, \text{Compress}(s_1))$.

Fig. 3.3 and Algorithm 2 sketches the signing procedure for FALCON and shows its pseudo-code for FALCON, respectively.

Readers can refer to Algorithm 11 in [9] for details on how to perform ffsampling. **SamplerZ** illustrated at Algorithm 15 in [9], for given inputs μ and σ' in a certain range, outputs an integer $z \sim D_{\mathbb{Z}, \sigma', \mu}$ in an isochronous manner.

3.3 Signing of FALCON

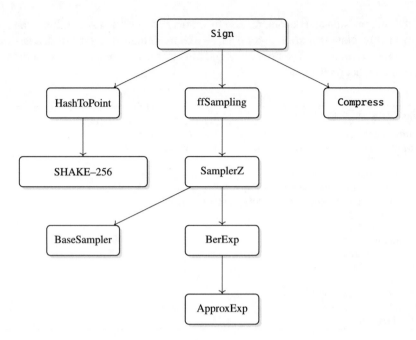

Fig. 3.3 Flowchart of Sign for FALCON

Algorithm 2 Sign of FALCON

Input: A message $M \in \{0,1\}^*$, secret key sk, a bound γ.
Output: A pair $(r, \text{Compress}(\mathbf{s}_1))$ with $r \in \{0,1\}^{320}$ and $\|(\mathbf{s}_1, \mathbf{s}_2)\| \leq \gamma$.
1: $r \leftarrow \mathcal{U}(\{0,1\}^{320})$
2: $\mathbf{c} \leftarrow \text{HashToPoint}(r\|M, q, n)$
3: $\mathbf{t} \leftarrow (-\frac{1}{q}\text{FFT}(c) \odot \text{FFT}(F), \frac{1}{q}\text{FFT}(c) \odot \text{FFT}(f))$ // $\mathbf{t} = (\text{FFT}(c), \text{FFT}(0)) \cdot \hat{\mathbf{B}}^{-1}$
4: **do**
5:     **do**
6:         $\mathbf{z} \leftarrow \text{ffSampling}_n(\mathbf{t}, T)$
7:         $\mathbf{s} = (\mathbf{t} - \mathbf{z})\hat{\mathbf{B}}$ // At this point, $\mathbf{s}$ follows Gaussian distribution.
8:     **while** $\|s\|^2 > \gamma$
9:     $(s_1, s_2) \leftarrow \text{FFT}^{-1}(\mathbf{s})$
10:     $s \leftarrow \text{Compress}(s_2, 8 \cdot \text{sbytelen} - 328)$ // Remove 1 byte for the header, and 40 bytes for r
11: **while**$(s = \bot)$
12: **return** (r, s)

3.3.1 Compress and Decompress Algorithms

The specification [9] of FALCON suggests encoding and decoding algorithms to reduce the size of keys and signatures. For completeness, we provide a description of the compression and decompression functions as depicted in Algorithms 3 and 4, respectively. Note that $slen = 8 \cdot |sgn| - 320$ by default where $|sgn|$ denotes the signature size in bytes. The `Compress` and `Decompress` techniques are generic and have no impact on the security level.

Algorithm 3 Compress

Input: A polynomial $s = \sum_{i=0}^{d-1} s_i X^i \in R = \mathbb{Z}[X]/(X^d + 1)$ and an integer $slen$.
Output: A compressed representation of str of s of bitsize $slen$, or $\bot$.
1: $str \leftarrow \{\}$
2: **for** $i = 0$ to $d - 1$ **do**
3: $\quad str \leftarrow (str \parallel b)$ where $b = 1$ if $s_i < 0$, $b = 0$ otherwise;
4: $\quad str \leftarrow (str \parallel b_6 b_5 \cdots b_0)$ where $b_j = (|s_i| \gg j)\&0x1$;
5: $\quad k \leftarrow |s_i| \gg 7$;
6: $\quad str \leftarrow (str \parallel 0^k 1)$
7: **end for**
8: **if** $|str| > slen$ **then**
9: $\quad str \leftarrow \bot$;
10: **else**
11: $\quad str \leftarrow (str \parallel 0^{slen-|str|})$
12: **end if**
13: **return** str

Algorithm 4 Decompress

Input: A bitstring str of bitsize $slen$
Output: A polynomial $s = \sum_{i=0}^{d-1} s_i X^i \in R = \mathbb{Z}[X]/(X^d + 1)$ or $\bot$
1: **if** $|str| \neq slen$ **then**
2: $\quad$ **return** $\bot$;
3: **end if**
4: **for** $i = 0$ to $d - 1$ **do**
5: $\quad s'_i \leftarrow \sum_{j=0}^{6} 2^{6-j} str[1 + j]$;
6: $\quad k \leftarrow 0$;
7: $\quad$ **while** $str[8 + k] = 0$ **do**
8: $\quad\quad k \leftarrow k + 1$
9: $\quad$ **end while**
10: $\quad s_i \leftarrow (-1)^{str[0]} \cdot (s'_i + 2^7 k)$;
11: $\quad$ **if** $s_i = 0$ and $str[0] = 1$ **then**
12: $\quad\quad$ **return** $\bot$
13: $\quad$ **end if**
14: $\quad str \leftarrow str[9 + k :]$
15: **end for**
16: **if** $|str| \neq 0^{|str|}$ **then**
17: $\quad$ **return** $\bot$;
18: **end if**
19: **return** $s = \sum_{i=0}^{d-1} s_i X^i$

3.4 Verification of FALCON

The last step of the scheme is thankfully simpler to describe. Upon receiving a signature $(r, \mathbf{s})$ and message M, the verifier decompresses $\mathbf{s}$ to a polynomial $\mathbf{s}_1$ and $\mathbf{c} = (0, \mathrm{H}(r||M))$, then wants to recover the full signature vector $\mathbf{v} = (\mathbf{s}_1, \mathbf{s}_2)$. If $\mathbf{v}$ is a valid signature, the verification identity is $(h, -1) \cdot (\mathbf{c} - \mathbf{v}) = -\mathrm{H}(r||M) - h\mathbf{s}_1 + \mathbf{s}_2 \bmod q = 0$, or equivalently the verifier can compute

$$\mathbf{s}_2 = \mathrm{H}(r||M) + h\mathbf{s}_1 \bmod q.$$

This is computed in the ring R_q, and can be done very efficiently for a good choice of modulus q using the Number Theoretic Transform (NTT). FALCON currently follow the standard choice of $q = 12{,}289$, as the multiplication in NTT format amounts to d integer multiplications in $\mathbb{Z}/q\mathbb{Z}$. The last step is to check that $\|(\mathbf{s}_1, \mathbf{s}_2)\|^2 \leq \gamma^2$: the signature is only accepted in this case. The rejection bound γ comes from the expected length of vectors outputted by Sample described as Algorithm 4 in [24].

Since they are morally Gaussian, they concentrate around their standard deviation; a "slack" parameter $\tau = 1.042$ is tuned to ensure that 90% of the vectors generated by Sample will get through the loop:

$$\gamma = \tau \cdot \sigma_{\mathrm{sig}} \cdot \sqrt{2d}.$$

Algorithm 5 shows the pseudo-code of verification procedure of FALCON.

Algorithm 5 Verif of FALCON

Input: A signature $(r, \mathbf{s})$ on M, a public key $\mathrm{pk} = h$, a bound γ.
Output: Accept or Reject.
1: $\mathbf{s}_1 \leftarrow$ Decompress($\mathbf{s}$)
2: $\mathbf{c} \leftarrow \mathrm{H}(r||M)$
3: $\mathbf{s}_2 \leftarrow \mathbf{c} + h\mathbf{s}_1 \bmod q$
4:
5: **if** $\|(\mathbf{s}_1, \mathbf{s}_2)\|^2 > \gamma^2$ **then**
6:     **return** Reject.
7: **else**
8:     **return** Accept.
9: **end if**

Chapter 4
SOLMAE Algorithm

4.1 Overview

Inspired by FALCON's design, Espitau et al. presented so-called MITAKA [6] to reduce some drawbacks of FALCON. At a high-level, it removes the inherent technicality of the sampling procedure, and most of its induced complexity from an implementation standpoint, for *free*, that is with no loss of efficiency. The simplicity of our design translates into faster operations while preserving signature and verification key sizes, in addition to allowing for additional features absent from FALCON, such as enjoying less expensive masking, and being parallelizable. In 2023, Espitau et al. [8] suggested so-called ANTRAG in order to improve MITAKA without loss of security covering all NIST level of security I to V using the degree of cyclotomic ring from 512 to 1024 over specific cyclotomic polynomials under the prime modulus but is not limited to the power of 2.

Taking all advantages of FALCON, MITAKA and ANTRAG, SOLMAE is yet another quantum-safe signature based on NTRU trapdoor and achieves *better performance* for the *same security and advantages* as FALCON which focused only on NIST I and V levels of security. More precisely, SOLMAE offers the "best of three worlds" between FALCON, MITAKA and ANTRAG. Overall, SOLMAE is summarized in Fig. 4.1. For details on SOLMAE, refer to [24].

More details about all the objects mentioned in this section can be found later. Here, we focus on the big lines behind our scheme's principles, keeping details at a minimum. While its predecessor FALCON could be summed up as "an efficient instantiation of the GPV framework", SOLMAE takes it one step further. The ingredients behind the boxes in Fig. 4.1 are as follows:

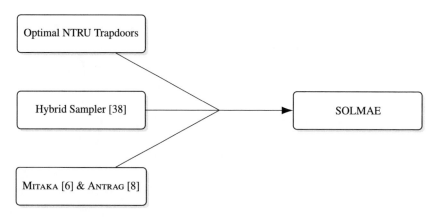

Fig. 4.1 Overview of SOLMAE

- An **optimally tuned key generation algorithm**, enhancing the security of our new sampler to that of FALCON's level;[1]
- The **hybrid sampler** is a faster, simpler, parallelizable and maskable Gaussian sampler to generate signatures;
- **Easy implementation** by assembling all the advantages of Mitaka and Antrag to make faster and simpler for practical purposes.

On the other hand, other techniques require tweaking the key generation and signing procedures.

4.2 Key Generation of SOLMAE

An important concern here is that not all pairs (f, g), (F, G) gives good trapdoor pairs for Sample described as Algorithm 4 in [24]. Schemes such as FALCON and Mitaka solve this technicality essentially by sieving among all possible bases to find the ones that reach an acceptable quality for the Sample procedure. This technique is costly, and many tricks were used to achieve an acceptable KeyGen. *This sieving routine was bypassed by redesigning completely how good quality bases can be found.* This improves the running time of KeyGen and also increases the security offered by Sample. In any case, note that NtruSolve's running time largely dominates the overall time for KeyGen: this is not avoidable as the basis completion algorithm requires working with quite large integers and relatively high-precision floating-point arithmetic.

At the end of the procedure, the secret key contains not only the secret basis but also the necessary data for Sign and Sample. This additional information can be

---

[1] This corresponds to the NIST-I and NIST-V requirements.

4.2 Key Generation of SOLMAE

represented by elements in $K_\mathbb{R}$ and is computed during or at the end of `NtruSolve`. All-in-all, `KeyGen` outputs:

$$\text{sk} = (\mathbf{b}_1 = (f, g), \mathbf{b}_2 = (F, G), \widetilde{\mathbf{b}}_2 = (\widetilde{F}, \widetilde{G}), \Sigma_1, \Sigma_2, \beta_1, \beta_2)),$$
$$\text{pk} = (h, q, \sigma_{\text{sig}}, \eta),$$

where we recall that $h = g/f \mod q$. These parameters and a table of their practical values are described more thoroughly in [24].

Informally, they correspond to the following:

- $(f, g), (F, G)$ is a good basis of the lattice $\mathcal{L}_{\text{NTRU}}$ associated to h, with quality $Q(f, g) = \alpha$, and $\widetilde{\mathbf{b}}_2$ is the Gram-Schmidt orthogonalization of (F, G) with respect to (f, g);
- $\sigma_{\text{sig}}, \eta$ are respectively the standard deviation for signature vectors, and a tight upper bound on the "smoothing parameter of $\mathbb{Z}^d$";
- $\Sigma_1, \Sigma_2 \in K_\mathbb{R}$ represent covariance matrices for two intermediate Gaussian samplings in `Sample`;
- the vectors $\beta_1, \beta_2 \in K_\mathbb{R}^2$ represent the orthogonal projections from $K_\mathbb{R}^2$ onto $K_\mathbb{R} \cdot \mathbf{b}_1$ and $K_\mathbb{R} \cdot \mathbf{b}_2$ respectively. In other words, they act as "getCoordinates" for vectors in $K_\mathbb{R}^2$. They are used by `Sample` and are precomputed for efficiency.

Algorithm 6 computes the necessary data for signature sampling, then outputs the key pair. Note that `NtruSolve` could also compute the sampling data and the public key, but for clarity, the pseudo-code gives these tasks to `KeyGen` of SOLMAE. Fig. 4.2 sketches the key generation procedure of SOLMAE.

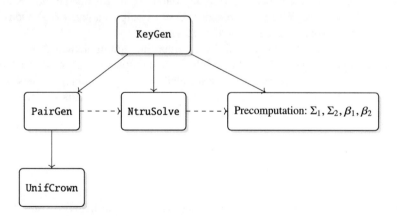

Fig. 4.2 Flowchart of KeyGen for SOLMAE

Algorithm 6 KeyGen of SOLMAE

Input: A modulus q, a target quality parameter $1 < \alpha$, parameters $\sigma_{\text{sig}}, \eta > 0$
Output: A basis $((f, g), (F, G)) \in R^2$ of an NTRU lattice $\mathcal{L}_{\text{NTRU}}$ with $Q(f, g) = \alpha$;
1: **while** f is invertible modulo q **do**
2:     $\mathbf{b}_1 := (f, g) \leftarrow \text{PairGen}(q, \alpha, R_-, R_+)$ // Secret basis computation between R_- and R_+
3: **end while**
4: $\mathbf{b}_2 := (F, G) \leftarrow \text{NtruSolve}(q, f, g)$:
5: $h \leftarrow g/f \mod q$ // Public key data computation
6: $\gamma \leftarrow 1.1 \cdot \sigma_{\text{sig}} \cdot \sqrt{2d}$ // tolerance for signature length
7: $\beta_1 \leftarrow \frac{1}{\langle \mathbf{b}_1, \mathbf{b}_1 \rangle_K} \cdot \mathbf{b}_1$ // Sampling data computation, in Fourier domain
8: $\Sigma_1 \leftarrow \sqrt{\frac{\sigma_{\text{sig}}^2}{\langle \mathbf{b}_1, \mathbf{b}_1 \rangle_K} - \eta^2}$
9: $\tilde{\mathbf{b}}_2 := (\tilde{F}, \tilde{G}) \leftarrow \mathbf{b}_2 - \langle \beta_1, \mathbf{b}_2 \rangle \cdot \mathbf{b}_1$
10: $\beta_2 \leftarrow \frac{1}{\langle \tilde{\mathbf{b}}_2, \tilde{\mathbf{b}}_2 \rangle_K} \cdot \tilde{\mathbf{b}}_2$
11: $\Sigma_2 \leftarrow \sqrt{\frac{\sigma_{\text{sig}}^2}{\langle \tilde{\mathbf{b}}_2, \tilde{\mathbf{b}}_2 \rangle_K} - \eta^2}$
12: $\text{sk} \leftarrow (\mathbf{b}_1, \mathbf{b}_2, \tilde{\mathbf{b}}_2, \Sigma_1, \Sigma_2, \beta_1, \beta_2)$
13: $\text{pk} \leftarrow (q, h, \sigma_{\text{sig}}, \eta, \gamma)$
14: **return** sk, pk

The function of two subroutines `PairGen` and `NtruSolve` are described below:

1. The `PairGen` algorithm generates d complex numbers $(x_j e^{i\theta_j})_{j \leq d/2}$, $(y_j e^{i\theta_j})_{j \leq d/2}$ to act as the FFT representations of two *real* polynomial $f^{\mathbb{R}}, g^{\mathbb{R}}$ in $K_{\mathbb{R}}$. The magnitude of these complex numbers is sampled in a planar annulus whose small and big radii are set to match a target $Q(f, g)$ with `UnifCrown` [24]. It then finds close elements $f, g \in \mathcal{Z}$ by round-off, unless maybe the rounding error was too large. When the procedure ends, it outputs a pair (f, g) such that $Q(f, g) = \alpha$, where α depends on the security level.

2. `NtruSolve` is exactly Pornin and Prest's algorithm and implementation [38]. It takes as input $(f, g) \in \mathcal{Z}^2$ and a modulus q, and outputs $(F, G) \in \mathcal{Z}^2$ such that $(f, g), (F, G)$ is a basis of $\mathcal{L}_{\text{NTRU}}$ associated to $h = g/f \mod q$. It does so by solving the Bézout-like equation $fG - gF = q$ in $\mathcal{Z}$ using recursively the tower of subfields for optimal efficiency.

4.3 Signing of SOLMAE

Recall that NTRU lattices live in $\mathbb{R}^{2d}$. Their structure also helps to simplify the preimage computation. Indeed, the signer only needs to compute $\mathbf{m} = H(M) \in \mathbb{R}^d$, as then $\mathbf{c} = (0, \mathbf{m})$ is a valid preimage: the corresponding polynomials satisfy $(h, 1) \cdot \mathbf{c} = \mathbf{m}$.

As the same with `Sign` procedure of FALCON, an interesting feature is that only the *first half* of the signature $(\mathbf{s}_1, \mathbf{s}_2) \in \mathcal{L}_{\text{NTRU}}$ needs to be sent along the message,

4.3 Signing of SOLMAE

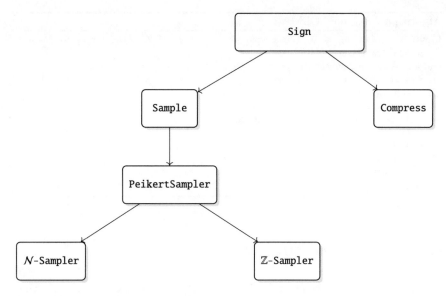

Fig. 4.3 Flowchart of Sign for SOLMAE

as long as h is available to the verifier. This comes from the identity $hs_1 = s_2 \mod q$ defining these lattices, as we will see in the Verif algorithm description.[2]

Because of their nature as Gaussian integer vectors, signatures can be encoded to reduce the size of their bit-representation. The standard deviation of Sample is large enough so that the $\lfloor \log \sqrt{q} \rfloor$ least significant bits of one coordinate are essentially random.

In practice, Sign adds a random "salt" $r \in \{0, 1\}^k$, where k is large enough that an unfortunate collision of messages is unlikely to happen, that is, it hashes $(r||M)$ instead of M—our analysis in this regard is identical to FALCON. A signature is then $\text{sig} = (r, \text{Compress}(s_1))$ using Algorithm 3 stated in Sect. 3.3.1. SOLMAE cannot output two different signatures for a message like FALCON which was mentioned in Sect. 3.3.

Fig. 4.3 sketches the signing procedure of SOLMAE and Algorithm 7 shows its pseudo-code. $\mathbb{Z}$-Sampler is equivalent to **SamplerZ** used in FALCON. $\mathcal{N}$-Sampler (Algorithm 10 in [24]) refers to sampling from a Gaussian or normal distribution. For PeikertSampler, see Algorithm 5 in [24].

---

[2] The same identity can also be used to check the validity of signatures only with a hash of the public key h, requiring this time send both s_1 and s_2, but we will not consider this setting here.

Algorithm 7 Sign of SOLMAE

Input: A message $M \in \{0,1\}^*$, a tuple $\mathsf{sk} = ((f,g),(F,G),(\widetilde{F},\widetilde{G}),\sigma_{\mathsf{sig}},\Sigma_1,\Sigma_2,\eta)$, a rejection parameter $\gamma > 0$.
Output: A pair $(r, \mathtt{Compress}(\mathbf{s}_1))$ with $r \in \{0,1\}^{320}$ and $\|(\mathbf{s}_1,\mathbf{s}_2)\| \leq \gamma$.
1: $r \leftarrow \mathcal{U}(\{0,1\}^{320})$
2: $\mathbf{c} \leftarrow (0, \mathsf{H}(r\|M))$
3: $\hat{\mathbf{c}} \leftarrow \mathsf{FFT}(\mathbf{c})$
4: **while** $\|(\mathsf{FFT}^{-1}(\hat{s}_1), \mathsf{FFT}^{-1}(\hat{s}_2))\|^2 \leq \gamma^2$ **do**
5: $\quad (\hat{s}_1, \hat{s}_2) \leftarrow \hat{\mathbf{c}} - \mathtt{Sample}(\hat{\mathbf{c}}, \mathsf{sk})$ $\quad\quad$ // $(\mathbf{s}_1,\mathbf{s}_2) \leftarrow D_{\mathcal{L}_{\mathsf{NTRU}},\mathbf{c},\sigma_{\mathsf{sig}}}$
6: **end while**
7: $s_1 \leftarrow \mathsf{FFT}^{-1}(\hat{s}_1)$
8: $s \leftarrow \mathtt{Compress}(s_1)$
9: **return** (r, s)

4.4 Verification of SOLMAE

The last step of the scheme is thankfully simpler to describe as shown in Algorithm 8. Upon receiving a signature (r, s) and message M, the verifier decompresses s to a polynomial s_1 and $\mathbf{c} = (0, \mathsf{H}(r\|M))$ to recover the full signature vector $\mathbf{v} = (s_1, s_2)$. If $\mathbf{v}$ is a valid signature, the verification identity is $(h, -1) \cdot (\mathbf{c} - \mathbf{v}) = -\mathsf{H}(r\|M) - hs_1 + s_2 \mod q = 0$, or equivalently the verifier can compute

$$s_2 = \mathsf{H}(r\|M) + hs_1 \mod q.$$

This is computed in the ring R_q, and can be performed very efficiently for a good choice of modulus q using the Number Theoretic Transform (NTT). We currently follow the standard choice (as in FALCON) of $q = 12,289$, as the multiplication in NTT format amounts to d integer multiplications in $\mathbb{Z}/q\mathbb{Z}$. The last step is to check that $\|(\mathbf{s}_1, \mathbf{s}_2)\|^2 \leq \gamma^2$: the signature is only accepted in this case.

The rejection bound γ comes from the expected length of vectors outputted by Sample. Since they are morally Gaussian, they concentrate around their standard deviation; a "slack" parameter $\tau = 1.042$ is tuned to ensure that 90% of the vectors generated by Sample will pass through the loop:

$$\gamma = \tau \cdot \sigma_{\mathsf{sig}} \cdot \sqrt{2d}.$$

4.4 Verification of SOLMAE

Algorithm 8 Verif of SOLMAE

Input: A signature (r, s) on M, a public key $\text{pk} = h$, a bound γ.
Output: Accept or reject.
1: $s_1 \leftarrow \texttt{Decompress}(s)$;
2: $c \leftarrow \text{H}(r||M)$;
3: $s_2 \leftarrow c + hs_1 \bmod q$;
4: **if** $\|(\mathbf{s}_1, \mathbf{s}_2)\|^2 > \gamma^2$ **then**
5:     **return** Reject.
6: **else**
7:     **return** Accept.
8: **end if**

Chapter 5
Basics of Python

5.1 Python Programming Language

Python Programming Language, simply Python, is a high-level, interpreted programming language known for its simplicity and readability. Created by Guido van Rossum in the late 1980s, Python has become one of the most popular programming languages, used across various domains, including web development, scientific computing, data analysis, artificial intelligence, and more.

1. Key Features:

   - Clear and Readable Syntax: Python's simple, easy-to-learn syntax emphasizes readability, making it an excellent choice for beginners and reducing the cost of program maintenance.
   - Dynamic Typing and Binding: These features allow Python to be highly flexible and suitable for rapid application development.
   - Modularity and Code Reuse: Python supports modules and packages, encouraging program modularity and code reuse.
   - Extensive Standard Library: Python comes with a vast standard library, providing tools and functionalities for virtually every task, with additional third-party modules available.
   - Cross-Platform: Python code can run on different platforms without modification, making it highly portable. Multiple Programming Paradigms: Python supports object-oriented, procedural, and functional programming styles, allowing developers to choose the best approach for their project.
   - Exception-Based Error Handling: Python handles errors by raising exceptions, which makes debugging easier. A source-level debugger allows inspection of variables, setting breakpoints, and stepping through code.

- Embeddable: Python can be embedded within applications as a scripting interface, adding flexibility to larger systems. Extensibility: Python can be extended with modules written in C, C++, Java (for Jython), or .NET languages (for IronPython).
2. Productivity: Python's fast edit-test-debug cycle enhances productivity. Since there is no compilation step, developers can quickly iterate on their code. Debugging is straightforward, and Python's introspective capabilities allow developers to inspect and manipulate objects at runtime.
3. Popular Frameworks and Libraries:
   - Web Development: Flask and Django are popular frameworks for building web applications.
   - Scientific Computing and Data Analysis: NumPy and Pandas are widely used for data manipulation and analysis.
   - Artificial Intelligence and Machine Learning: Libraries like TensorFlow, Keras, and PyTorch make Python a popular choice in AI/ML domains.

In conclusion, Python's simplicity, flexibility, and extensive library support make it a powerful language for a wide range of tasks, appealing to both beginners and experienced developers alike.

5.2 Python Environment for Windows OS

You can download Visual Studio Code (or your favourable IDE application) for your platform-Windows OS 10 or 11, Debian, Ubuntu, or macOS 10.15+—from https://code.visualstudio.com/download. The installation process is straightforward; simply follow the provided instructions. Similarly to set up Visual Studio Code, unzip the `VSCode-win32-arm64-1.92.2.zip` archive (unpacked size: 391,456,013 bytes) and install it in your desired folder at your computer. It is convenient to install WSL (Windows Subsystem for Linux) at your personal computer working Windows OS to verify your program working Windows and Unix OS's at one platform. WSL allows developers to use Linux command-line tools and utilities alongside Windows applications, making it easier to work in a Linux environment for tasks such as software development, system administration, and more, without leaving Windows ecosystem. It is better to install WSL 2.0 which was introduced in 2019 and offers better compatibility with Linux software,

5.3 Useful Python Packages

Fig. 5.1 Screen capture of Visual Studio Code

faster file I/O performance, and full system call compatibility. For Unix or MacOS platform, it is also easy to set up Python environment. Fig. 5.1 shows the execution screen of Visual Studio Code on Windows 10. The leftmost window displays the file folder information, the upper-right window shows the editing screen for the selected executable file, and the lower-right window displays the execution terminal. In this screen, the file `test_merge_and_split.py` is being run with Python 3.8.9, and the lower-right window shows the execution results by displaying its outcome of five test cases. Its details are described at Sect. 6.1.1. This setup makes it incredibly convenient to edit and execute your Python scripts directly on your personal computer.

5.3 Useful Python Packages

To view the necessary packages installed, use the command `pip install list` in Microsoft Visual Studio Code, which will display the installed packages, as shown in Fig. 5.2. For example, you may need to install packages such as **numpy**.

```
Package                    Version
------------------------   ---------
certifi                    2022.12.7
charset-normalizer         2.1.1
contourpy                  1.0.6
cycler                     0.11.0
ecdsa                      0.19.0
fonttools                  4.38.0
idna                       3.4
kiwisolver                 1.4.4
matplotlib                 3.6.2
Naked                      0.1.32
numpy                      1.24.1
packaging                  22.0
pandas                     1.5.3
Pillow                     9.3.0
pip                        22.3.1
pycryptodome               3.16.0
pycryptodome-test-vectors  1.0.11
pyparsing                  3.0.9
python-dateutil            2.8.2
pytz                       2022.7.1
PyYAML                     6.0
requests                   2.28.1
scipy                      1.10.1
setuptools                 49.2.1
shellescape                3.8.1
six                        1.16.0
urllib3                    1.26.13
wheel                      0.41.2
```

Fig. 5.2 Installed packages in my PC environment

Chapter 6
Checking FALCON with Python

https://github.com/tprest/falcon.py contains an implementation of the FALCON post-quantum cryptographic signature scheme in Python at github repository. This repository contains the following files (roughly in order of dependency):

1. common.py contains shared functions and constants
2. encoding.py contains compression and decompression
3. rng.py implements a ChaCha20-based PRNG, useful for KATs (standalone)
4. samplerz.py implements a Gaussian sampler over the integers (standalone)
5. fft_constants.py contains precomputed constants used in the FFT
6. ntt_constants.py contains precomputed constants used in the NTT
7. fft.py implements the FFT over $R[x]/(x^n + 1)$
8. ntt.py implements the NTT over $Z_q[x]/(x^n + 1)$
9. ntrugen.py generate polynomials f, g, F, G in $Z[x]/(x^n + 1)$ such that $f \cdot G - g \cdot F = q$
10. ffsampling.py implements the fast Fourier sampling algorithm
11. falcon.py implements FALCON
12. test.py implements tests to check that everything is properly implemented

Under ..\scripts folder contains some files that are helpful to implement FALCON, test it and understand where parameters/constants come from. This repository contains the following files:

1. generate_constants.sage can be used in SageMath to generate the FFT and NTT constants.
2. parameters.py is a script that generates parameters used in the Round 3 specification as well as the C implementation.
3. saga.py contains the SAGA (Statistically Acceptable GAussians) test [17] suite to test Gaussian samplers. It is used in test.py.
4. samplerz_KAT512.py and samplerz_KAT1024.py contain test vectors for the sampler over the integers. They are used in test.py.

5. sign_KAT.py contains test vectors for the signing procedure. It is used in test.py.

To execute generate_constants.sage, familiarity with the SageMath programming language (https://www.sagemath.org) is required, which is beyond the scope of this monograph. For further details, please consult the appropriate references.

6.1 Utility Modules for FALCON

6.1.1 Checking common.py

The Python script in Script 6.1 is commonly used in other Python scripts when implementing FALCON or SOLMAE. The modular value q is fixed at 12,289 but can be adjusted based on your specific application. Since FALCON employs polynomial arithmetic, the script enables splitting a polynomial into two equal parts and merging them back into a single polynomial, as shown in Script 6.1. The sqnorm(v) definition in common.py is not required for this test.

```python
"""Contains methods and objects which are reused through
    multiple files."""
"""q is the integer modulus which is used in Falcon."""
q = 12 * 1024 + 1
def split(f):
    """Split a polynomial f in two polynomials.
    Args:
        f: a polynomial
    Format: coefficient
    """
    n = len(f)
    f0 = [f[2 * i + 0] for i in range(n // 2)]
    f1 = [f[2 * i + 1] for i in range(n // 2)]
    return [f0, f1]
def merge(f_list):
    """Merge two polynomials into a single polynomial f."""
    Args:
        f_list: a list of polynomials
    Format: coefficient
    """
    f0, f1 = f_list
    n = 2 * len(f0)
    f = [0] * n
    for i in range(n // 2):
        f[2 * i + 0] = f0[i]
        f[2 * i + 1] = f1[i]
    return f
def sqnorm(v):
    """Compute the square euclidean norm of the vector v."""
```

6.1 Utility Modules for FALCON

```
29      res = 0
30      for elt in v:
31          for coef in elt:
32              res += coef ** 2
33      return res
```

Script 6.1 Script of test_split_and_merge()

To verify the correctness of the split(f) and merge(f_list) functions shown in Script 6.1, you can run the test Python script provided in Script 6.2. This script checks four test cases with varying polynomial degrees to ensure proper functionality. The Python script can be executed in your environment as test_split_and_merge() function runs automatically under
if __name__ == "__main__":.

```
1   def test_split_and_merge():
2       # Test case 1: Polynomial with even number of coefficients
3       f1 = [3, 2, 1, 4, 5, 6]  # Represents 3x^5 + 2x^4 + 1x^3 +
            4x^2 + 5x +6
4       split_f1 = split(f1)
5       merged_f1 = merge(split_f1)
6       print("Test 1 - Original:", f1, "-> Split:", split_f1, "->
            Merged:", merged_f1)
7       # Check if merged is equal to original
8       assert f1 == merged_f1, "Test 1 failed: Merged one does not
            match the original"
9       # Test case 2: Polynomial with odd number of coefficients
10      f2 = [4, 2, 5, 7, 9, 11]  # Represents 4x^5 + 2X^4+ 5x^3 +
            7x^2 + 9x + 11
11      split_f2 = split(f2)
12      merged_f2 = merge(split_f2)
13      print("Test 2 - Original:", f2, "-> Split:", split_f2, "->
            Merged:", merged_f2)
14      # Check if merged is equal to original
15      assert f2 == merged_f2, "Test 2 failed: Merged one does not
            match the original"
16      # Test case 3: Polynomial with minimal length
17      f3 = [1, 2]  # Represents 1x + 2
18      split_f3 = split(f3)
19      merged_f3 = merge(split_f3)
20      print("Test 3 - Original:", f3, "-> Split:", split_f3, "->
            Merged:", merged_f3)
21      # Check if merged is equal to original
22      assert f3 == merged_f3, "Test 3 failed: Merged one does not
            match the original"
23      # Test case 4: Empty polynomial
24      f4 = []  # Represents an empty polynomial
25      split_f4 = split(f4)
26      merged_f4 = merge(split_f4)
27      print("Test 4 - Original:", f4, "-> Split:", split_f4, "->
            Merged:", merged_f4)
28      # Check if merged is equal to original
```

```
29      assert f4 == merged_f4, "Test 4 failed: Merged one does not
            match the original"
30      print("All tests passed!")
31  if __name__ =="__main__":
32  # Run the test case function
33      test_split_and_merge()
```

Script 6.2 test_split_and_merge.py

```
Test 1 - Original: [3, 2, 1, 4, 5, 6] -> Split: [[3, 1, 5], [2, 4, 6]]
                                      -> Merged: [3, 2, 1, 4, 5, 6]
Test 2 - Original: [4, 2, 5, 7, 9, 11] -> Split: [[4, 5, 9], [2, 7, 11]]
                                       -> Merged: [4, 2, 5, 7, 9, 11]
Test 3 - Original: [1, 2] -> Split: [[1], [2]] -> Merged: [1, 2]
Test 4 - Original: [] -> Split: [[], []] -> Merged: []
All tests passed!
```

Fig. 6.1 Output of test_split_and_merge()

After executing test_split_and_merge_test.py as shown in Script 6.2, the result obtained are displayed in Fig. 6.1. These results indicate that all four cases are working correctly by splitting and merging a given polynomial.

6.1.2 Checking fft.py

In order to perform FFT (Fast Fourier Transform) and IFFT (Inverse Fast Fourier Transform), which reduce the complexity of polynomial multiplication from $O(n^2)$, as in the schoolbook method, to $O(n \log n)$, we need to split and merge polynomials with real and complex values shown in Script 6.3 that we have once finished to check before.

```
1  def split_fft(f_fft):
2      """Split a polynomial f in two polynomials.
3      Args:
4          f: a polynomial
5      Format: FFT
6      Corresponds to algorithm 1 (splitfft_2) of Falcon's
            documentation.
7      """
8      n = len(f_fft)
9      w = roots_dict[n]
10     f0_fft = [0] * (n // 2)
11     f1_fft = [0] * (n // 2)
12     for i in range(n // 2):
13         f0_fft[i] = 0.5 * (f_fft[2 * i] + f_fft[2 * i + 1])
14         f1_fft[i] = 0.5 * (f_fft[2 * i] - f_fft[2 * i + 1]) *
                w[2 * i].conjugate()
```

6.1 Utility Modules for FALCON

```
        return [f0_fft, f1_fft]

def merge_fft(f_list_fft):
    """Merge two or three polynomials into a single polynomial f.
    Args:
        f_list: a list of polynomials
    Format: FFT
    Corresponds to algorithm 2 (mergefft_2) of Falcon's
        documentation.
    """
    f0_fft, f1_fft = f_list_fft
    n = 2 * len(f0_fft)
    w = roots_dict[n]
    f_fft = [0] * n
    for i in range(n // 2):
        f_fft[2 * i + 0] = f0_fft[i] + w[2 * i] * f1_fft[i]
        f_fft[2 * i + 1] = f0_fft[i] - w[2 * i] * f1_fft[i]
    return f_fft
```

Script 6.3 Functions of splitting and merging polynomials for fft.py

```
""" The roots of phi_16 = x^8 + 1
The second half of the roots are is the conjugates of the first half.
The root at index (2 * i + 1) is the negation of the root at index (2*i).
"""
phi16_roots = [0.923879532511287 + 0.382683432365090j,
              -0.923879532511287 - 0.382683432365090j,
               0.382683432365090 - 0.923879532511287j,
              -0.382683432365090 + 0.923879532511287j,
               0.923879532511287 - 0.382683432365090j,
              -0.923879532511287 + 0.382683432365090j,
               0.382683432365090 + 0.923879532511287j,
              -0.382683432365090 - 0.923879532511287j]
```

Fig. 6.2 phi16_roots used in Script 6.3 for FFT

To run Script 6.3 correctly, we need to call the computed list of the roots of phi_16 of x^8+1 for FFT which was precomputed by generate_constants.sage and stored at a part of fft_constants.py shown as Fig. 6.2.

Script 6.4 presents the FFT and IFFT functions, which use merge_fft(), split_fft(), and other related functions to convert real numbers in the time domain into complex numbers in the frequency domain recursively, and vice versa.

```
def fft(f):
    """Compute the FFT of a polynomial mod (x ** n + 1).
    Args:  f: a polynomial
    Format: input as coefficients, output as FFT
    """
    n = len(f)
    if (n > 2):
        f0, f1 = split(f)
```

```
            f0_fft = fft(f0)
            f1_fft = fft(f1)
            f_fft = merge_fft([f0_fft, f1_fft])
        elif (n == 2):
            f_fft = [0] * n
            f_fft[0] = f[0] + 1j * f[1]
            f_fft[1] = f[0] - 1j * f[1]
        return f_fft

def ifft(f_fft):
    """Compute the inverse FFT of a polynomial mod (x ** n + 1).
    Args:    f: a FFT of a polynomial
    Format:  input as FFT, output as coefficients
    """
    n = len(f_fft)
    if (n > 2):
        f0_fft, f1_fft = split_fft(f_fft)
        f0 = ifft(f0_fft)
        f1 = ifft(f1_fft)
        f = merge([f0, f1])
    elif (n == 2):
        f = [0] * n
        f[0] = f_fft[0].real
        f[1] = f_fft[0].imag
    return f
```

Script 6.4 Python functions used in fft.py

Script 6.5 presents a test program to verify the correct functioning of the FFT and IFFT implementations shown in Script 6.4.

```
from fft import fft,ifft
import numpy as np
def test_fft_ifft():
    # Define a polynomial of degree 8 (which has 9 coefficients)
    poly = [1, 2, 3, 4, 5, 6, 7, 8]
    print("\nBefore FFT:")
    print(np.array(poly))
    # Perform FFT
    fft_result = fft(poly)
    print("\nFFT Result:")
    print(np.array(fft_result))
    # Perform IFFT
    ifft_result = ifft(fft_result)
    print("\nIFFT Result:")
    print(np.array(ifft_result))
    # Check if IFFT result is close to original polynomial
    is_close = np.allclose(poly, ifft_result)
    print("\nTest passed:", is_close)
if __name__ =="__main__":
    # Run the test
    test_fft_ifft()
```

Script 6.5 Test program for fft.py

6.1 Utility Modules for FALCON

```
Before FFT:
[1 2 3 4 5 6 7 8]

FFT Result:
[-8.13707118+25.13669746j   4.48021694 -0.99456184j
  3.3800856  -7.48302881j   4.27676865 +3.34089319j
 -8.13707118-25.13669746j   4.48021694 +0.99456184j
  3.3800856  +7.48302881j   4.27676865 -3.34089319j]

IFFT Result:
[1. 2. 3. 4. 5. 6. 7. 8.]

Test passed: True
```

Fig. 6.3 Output of checking ftt.py

Fig. 6.3 presents an example output from the test program, demonstrating the correct functionality of both FFT and IFFT.

6.1.3 Checking ntt.py

Similarly to the FTT, the Number Theoretic Transform (NTT) and its inverse (INTT) perform various operations on polynomials in the number domain, utilizing modular arithmetic. The implementation details of these operations can be found in the ntt.py and ntt_constants.py files within the FALCON Python package. Script 6.6 illustrates the operation of the NTT and INTT, utilizing modular arithmetic. It also includes a test program to verify their correctness. Fig. 6.5 presents a test example to verify the correct functioning of the NTT and INTT.

```python
from common import split, merge, q
from ntt_constants import roots_dict_Zq, inv_mod_q
import numpy as np
i2 = 6145 # modular inverse of 2 for a given q
sqr1 = roots_dict_Zq[2][0]
def split_ntt(f_ntt):
    """Split a polynomial f in two or three polynomials."""
    n = len(f_ntt); w = roots_dict_Zq[n]
    f0_ntt = [0] * (n // 2); f1_ntt = [0] * (n // 2)
    for i in range(n // 2):
        f0_ntt[i] = (i2 * (f_ntt[2 * i] + f_ntt[2 * i + 1])) % q
        f1_ntt[i] = (i2 * (f_ntt[2 * i] - f_ntt[2 * i + 1]) *
                inv_mod_q[w[2 * i]]) % q
    return [f0_ntt, f1_ntt]
def merge_ntt(f_list_ntt):
    """Merge two or three polynomials into a single
        polynomial."""
    f0_ntt, f1_ntt = f_list_ntt
```

```python
        n = 2 * len(f0_ntt); w = roots_dict_Zq[n]
        f_ntt = [0] * n
        for i in range(n // 2):
            f_ntt[2 * i] = (f0_ntt[i] + w[2 * i] * f1_ntt[i]) % q
            f_ntt[2 * i + 1] = (f0_ntt[i] - w[2 * i] * f1_ntt[i]) % q
        return f_ntt
def ntt(f):
    """Compute the NTT of a polynomial."""
    n = len(f)
    if n > 2:
        f0, f1 = split(f)
        f0_ntt = ntt(f0); f1_ntt = ntt(f1)
        f_ntt = merge_ntt([f0_ntt, f1_ntt])
    elif n == 2:
        f_ntt = [0] * n
        f_ntt[0] = (f[0] + sqr1 * f[1]) % q
        f_ntt[1] = (f[0] - sqr1 * f[1]) % q
    return f_ntt
def intt(f_ntt):
    """Compute the inverse NTT of a polynomial."""
    n = len(f_ntt)
    if n > 2:
        f0_ntt, f1_ntt = split_ntt(f_ntt)
        f0 = intt(f0_ntt); f1 = intt(f1_ntt)
        f = merge([f0, f1])
    elif n == 2:
        f = [0] * n
        f[0] = (i2 * (f_ntt[0] + f_ntt[1])) % q
        f[1] = (i2 * inv_mod_q[1479] * (f_ntt[0] - f_ntt[1])) % q
    return f
def test_ntt_intt():
    """Test NTT and INTT functions for correctness."""
    import random
    n = 8      # Random polynomials of length 8 (a power of 2)
    f = [random.randint(0, q-1) for _ in range(n)]
    print("\nBefore NTT:"); print(np.array(f))
    # Compute NTT and then INTT
    f_ntt = ntt(f)
    print("\nNTT Result:"); print(np.array(f_ntt))
    f_intt = intt(f_ntt)
    print("\nINTT Result:"); print(np.array(f_intt))
    # Check if the INTT of NTT is the original polynomial
    assert f == f_intt, f"Test failed: {f} != {f_intt}"
    print(f"\nTest passed.")
if __name__ == "__main__":
# Run the test
    test_ntt_intt()
```

Script 6.6 Test script for ntt.py

To run Script 6.6 correctly, we need to call the computed list of the roots of phi_16 of x^8+1_Z for NTT which was precomputed by generate_constants.sage and stored at a part of ntt_constants.py shown as Fig. 6.4.

6.1 Utility Modules for FALCON

```
"""Roots of phi_16 = x^8 + 1_Z"""
phi16_roots_Zq = [5736, 6553, 4134, 8155, 722, 11567, 1305, 10984]
```

Fig. 6.4 phi16_roots used in Script 6.6 for NTT

```
Before NTT:
[12092  3897  6803   506  2551  6547  2140  1605]

NTT Result:
[ 5049  7760  9402  2403  2836  1273  2352  4216]

INTT Result:
[12092  3897  6803   506  2551  6547  2140  1605]

Test passed.
```

Fig. 6.5 Output of checking ntt.py

6.1.4 Checking ntrugen.py

To generate a private signing key, two randomly generated polynomials, f and g are required. From these polynomials, two additional polynomials, F and G are derived using the extended Euclidean algorithm under the modulus q. This process is implemented in the function ntrugen.py, which is provided in part of the FALCON Python package. To verify the correctness of ntrugen.py, we developed a Python test script as illustrated in Script 6.7. This step can sometimes be time-consuming.

```python
from common import q
from fft import sub
from ntrugen import karamul, ntru_gen
import numpy as np
def polynomial_mod(a, mod):
    n = len(mod)
    result = np.polydiv(a, mod)[1]
    return np.round(result).astype(int).tolist()
def check_ntru_properties(f, g, F, G, q):
    n = len(f)
    x_n_plus_1 = [0] * (n + 1)
    x_n_plus_1[0] = 1; x_n_plus_1[n] = 1
    fG = karamul(f, G); gF = karamul(g, F)
    fG_minus_gF = sub(fG, gF)
    print("fG_minus_gF:")
    print(np.array(fG_minus_gF))
    mod_result = polynomial_mod(fG_minus_gF, x_n_plus_1)
    q_poly = [q] + [0] * (n-1)
    return mod_result == q_poly
def test_ntru_gen(n):
    """Test the ntru_gen function."""
```

```
         # Generate polynomials
         f, g, F, G = ntru_gen(n)
         print("f:"); print(np.array(f))
         print("g:"); print(np.array(g))
         print("F:"); print(np.array(F));
         print("G:"); print(np.array(G));
         # Check if f and g are non-zero
         if all(coef == 0 for coef in f) or all(coef == 0 for coef in
             g):
             raise AssertionError("Generated polynomials f or g are
                 zero.")
         # Check if the NTRU property is satisfied
         if not check_ntru_properties(f, g, F, G, q):
             raise AssertionError("f * G - g * F = q mod(x^n+1) is
                 not satisfied.")
         print("NTRU generation test passed.")
if __name__ == "__main__":
         n = 16 # Fix polynomial degree to test
         test_ntru_gen(n)
```

Script 6.7 Test script for ntrugen.py

To verify the correctness of ntrugen.py, we set $n = 16$ and created a Python test script. This script, as shown in Fig. 6.6, generates an example set of the polynomials f, g, F, and G.

6.1.5 Checking encoding.py

This section explains the correctness of the compress and decompress functions used to reduce the memory footprint of signature and other data, if needed. The definitions of compress(v,slen) and decompress(x,slen,n) functions are already provided in encoding.py module of FALCON Python package. Script 6.8

```
f:
[ 10  -5   6  16  28  -4 -27  -8  39 -29  30  11 -11  25 -47  13]
g:
[ 47   1 -18  -4 -10 -29  -7  -9  25  14 -18  -5 -21  -2   6  11]
F:
[ -9 -19  -8 -45 -51  -9  12  15   0   6 -57 -13 -12  -1 -13  47]
G:
[  7  -6  51   0  36   9 -28  35 -58 -16  30  42 -52 -15 -56  -9]
fG_minus_gF:
[12289    0    0    0    0    0    0    0 .  0    0    0    0
      0    0    0    0]
NTRU generation test passed.
```

Fig. 6.6 Output of checking ntrugen.py

6.1 Utility Modules for FALCON

illustrates the 6 test cases coded to verify the correctness of compress(v,slen) and decompress(x,slen,n) functions when used together.

```python
from encoding import compress, decompress
def test_compress_decompress():
    # Test cases
    test_cases = [
        ([-128, 127, 0], 5),    # Mixed negative, positive, and
                                  zero values
        ([0, 0, 0, 0], 2),       # All zeros
        ([1, 2, 3, 4, 5, 6, 7], 3),   # Small positive integers
        ([-1, -2, -3, -4, -5, -6, -7], 3),   # Small negative
                                               integers
        ([255, -255, 128, -128], 6),  # Edge cases for low and
                                        high bits
        ([32767, -32768], 6),    # Large positive and negative
                                   integers
    ]
    for i, (v, slen) in enumerate(test_cases):
        print(f"Test case {i + 1}: v = {v}, slen = {slen}")
        compressed = compress(v, slen)
        if compressed is False:
            print("Compression failed (encoding too long)")
        else:
            decompressed = decompress(compressed, slen, len(v))
            if decompressed is False:
                print("Decompression failed (invalid encoding)")
            elif decompressed == v:
                print("Success! Decompressed list matches the
                    original list.")
            else:
                print("Failure! Decompressed list does not match
                    the original list.")
            print(f"Compressed data: {compressed}\nDecompressed
                data: {decompressed}")
        print("-" * 40)

if __name__ == "__main__":
    test_compress_decompress()
```

Script 6.8 Test program for encoding.py

Fig. 6.7 displays the output of the six test cases after running those illustrated in Script 6.8. Two test cases succeeded, but other four cases failed due to encoding too long error. In practice, it is possible to handle longer data for compression and decompression.

```
Test case 1: v = [-128, 127, 0], slen = 5
Success! Decompressed list matches the original list.
Compressed data: b'\x80_\xe0\x10\x00'
Decompressed data: [-128, 127, 0]
----------------------------------------
Test case 2: v = [0, 0, 0, 0], slen = 2
Compression failed (encoding too long)
----------------------------------------
Test case 3: v = [1, 2, 3, 4, 5, 6, 7], slen = 3
Compression failed (encoding too long)
----------------------------------------
Test case 4: v = [-1, -2, -3, -4, -5, -6, -7], slen = 3
Compression failed (encoding too long)
----------------------------------------
Test case 5: v = [255, -255, 128, -128], slen = 6
Success! Decompressed list matches the original list.
Compressed data: b'\x7f\x7f\xd0\x06\x01\x00'
Decompressed data: [255, -255, 128, -128]
Decompressed data: [255, -255, 128, -128]
----------------------------------------
Test case 6: v = [32767, -32768], slen = 6
Compression failed (encoding too long)
----------------------------------------
```

Fig. 6.7 Output of six test cases

6.2 FALCON-Specific Modules

This section outlines the specific functions and operations exclusive to both FALCON-512 and FALCON-1024.

6.2.1 Checking `parameters.py`

`parameters.py`, located under folder `..\scripts` in the FALCON Python package, is used to generate two crucial sets of security parameters for FALCON-512 and FALCON-1024. It focuses on the parameters, metrics, and security aspects of these schemes. This section outlines its performance and output after executing `parameters.py`.

Fig. 6.8 provides a summary of the key definitions related to the secure use of FALCON across three aspects—Parameters, Metrics and Security. All parameter values can be established during the setup phase and made available to authorized signers and verifiers. The values in parenthesis use the optimization in [2].

Fig. 6.9 illustrates the specific values of various parameters for both FALCON-512 and FALCON-1024. Note that the value of $beta$ and $beta^2$ are approximated to the nearest integer value less than their actual values.

6.2 FALCON-Specific Modules

```
Parameters:
=======
- The degree of the ring Z[x]/(x^n + 1) is n.
- The integer modulus is q.
- The Gram-Schmidt norm is gs_norm: std. dev. of its signatures is sigma.
- The minimal std dev for sampling over Z is sigmin.
- The maximal std dev for sampling over Z is sigmax.
- The tailcut rate for signatures is tailcut.
- Signatures are rejected whenever ||(s1, s2)||^2 > beta^2.

Metrics:
=====
- The maximal number of signing queries is nb_queries.
- The signing rejection rate is rejection_rate.
- The maximal size of signatures is sig_bytesize (HARDCODED).

Security:
=========
- The targeted security level is target_bitsec.
- For x in {keyrec, forgery} (i.e. key recovery or forgery):
  - The BKZ blocksize required to achieve x is x_blocksize.
  - The classic CoreSVP hardness of x is x_coresvp_c.
  - The quantum CoreSVP hardness of x is x_coresvp_q.
```

Fig. 6.8 Description of parameters

6.2.2 Checking samplerz.py

SamplerZ generates random integer according to a Gaussian distribution with the specified mean and standard deviation. It utilizes `Basesampler()`, `Berexp()` and other computations as outlined in Algorithm 15 in [9].

The test script for samplerz.py, shown in Script 6.9, verifies whether 500 generated random integers, with a given mean of 0 and a standard deviation of 2.0, conform to the ideal Gaussian distribution.

```
from samplerz import samplerz
import numpy as np
import matplotlib.pyplot as plt
from scipy.stats import norm

def test_samplerz():
    mu = 0.0      # Mean of the distribution
    sigma = 2.0   # Standard deviation
    sigmin = 1.5  # sigmin scaling factor (must be 1 < sigmin <
         sigma < MAX_SIGMA)

    # Run the sampler multiple times to generate samples
    samples = [samplerz(mu, sigma, sigmin) for _ in range(500)]
    # Print out the first few samples
```

```
n        = 512
Parameters:
q        = 12289
gs_norm  = 129.7012417056984
sigma    = 165.7366171829776
sigmin   = 1.2778336969128337
sigmax   = 1.8205
tailcut  = 1.1
beta     = 5833
beta^2   = 34034726

Metrics:
========
nb_queries     = 2^64
rejection_rate = 4.906444693922141e-05
sig_bytesize   = 666

Security:
=========
target_bitsec     = 128
keyrec_blocksize  = 458 (418)
keyrec_coresvp_c  = 133 (122)
keyrec_coresvp_q  = 121 (110)
forgery_blocksize = 411 (374)
forgery_coresvp_c = 120 (109)
forgery_coresvp_q = 108 (99)
-------------------------------------------
n        = 1024
Parameters:
==========
q        = 12289
gs_norm  = 129.7012417056984
sigma    = 168.38857144654395
sigmin   = 1.298280334344292
sigmax   = 1.8205
tailcut  = 1.1
beta     = 8382
beta^2   = 70265242

Metrics:
========
nb_queries     = 2^64
rejection_rate = 2.407319953451673e-09
sig_bytesize   = 1280

Security:
=========
target_bitsec     = 256
keyrec_blocksize  = 936 (869)
keyrec_coresvp_c  = 273 (253)
keyrec_coresvp_q  = 248 (230)
forgery_blocksize = 952 (884)
forgery_coresvp_c = 277 (258)
forgery_coresvp_q = 252 (234)
```

Fig. 6.9 Specific values of various parameters for both FALCON-512 and FALCON-1024

```python
    print("First 15 generated integers:  ", f"{samples[:15]}
        ...]")

    # Calculate sample mean and sample standard deviation
    sample_mean = np.mean(samples)
    sample_std = np.std(samples)

    # Generate a range of values for the theoretical Gaussian
        distribution
    x_values = np.linspace(min(samples), max(samples), 1000)

    # Calculate the Gaussian probability density function (PDF)
        for comparison
    gaussian_pdf = norm.pdf(x_values, loc=mu, scale=sigma)

    # Plot the histogram of the samples and overlay the
        theoretical Gaussian curve
    plt.figure(figsize=(8, 6))
    plt.hist(samples, bins=20, density=True, edgecolor='black',
        alpha=0.7, label='Sample Histogram')
    plt.plot(x_values, gaussian_pdf, 'r-', label='Theoretical
        Gaussian PDF', linewidth=2)
    plt.title('Comparison of Sample Distribution with Gaussian
        Distribution')
    plt.xlabel('Sample Value'); plt.ylabel('Density')
    plt.legend(); plt.grid(True); plt.show()

    # Print the sample mean and standard deviation
    print(f'Sample Mean: {sample_mean}')
    print(f'Sample Standard Deviation: {sample_std}')

if __name__ == "__main__":
    test_samplerz()
print("\nTest passed:")
```

Script 6.9 Test program for `samplerz.py`

Fig. 6.10 presents the first 10 generated integers following a Gaussian distribution, making it challenging to visually determine the underlying distribution.

To compare the generated random integers with the ideal Gaussian distribution, which has a mean of 0.21 and a standard deviation of 1.870, Fig. 6.11 illustrates the comparison. The generated values appear consistent with the Gaussian distribution for the specified parameters. Since each test generates different random integers, the

```
First 10 generated integers:  [1, -1, -1, 4, 3, 5, 3, 2, 0] ...]
Sample Mean: 0.21
Sample Standard Deviation: 1.8702673605663978

Test passed:
```

Fig. 6.10 Output of `test_samplerz`

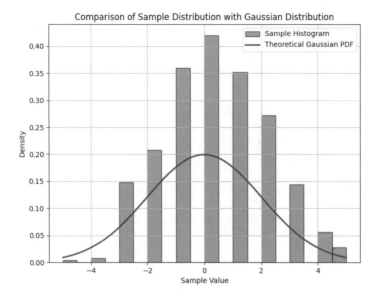

Fig. 6.11 Comparison of generated random integers with ideal Gaussian

figure will vary with each run. Sometimes the test fails. If this happens, retrying it might produce the desired result.

6.2.3 Checking `ffnp()` in `ffsampling.py`

This section describes to test the correctness of `ffnp()` function used in `ffsampling.py` which samples the random value close to the theoretical bound in a Fourier domain. This is unique idea used in FALCON. Script 6.10 shows its test script in Python.

```
"""
This file tests ffnp() function used at ffsampling.py in Falcon
    Python package.
"""
from common import q, sqnorm
from fft import add, sub, mul, div, neg, fft, ifft
from ffsampling import ffldl, ffldl_fft, ffnp, ffnp_fft
from ffsampling import gram
from random import randint, random, gauss, uniform
from ntrugen import karamul, ntru_gen, gs_norm
from scripts.sign_KAT import sign_KAT
def vecmatmul(t, B):
    """Compute the product t * B, where t is a vector and B is a
        square matrix.
    Args:
```

6.2 FALCON-Specific Modules

```
            B: a matrix
        Format: coefficient
        """
        nrows = len(B)
        ncols = len(B[0])
        deg = len(B[0][0])
        assert(len(t) == nrows)
        v = [[0 for k in range(deg)] for j in range(ncols)]
        for j in range(ncols):
            for i in range(nrows):
                v[j] = add(v[j], mul(t[i], B[i][j]))
        return v
def test_ffnp(n, iterations):
    """Test ffnp.
    This functions check that:
    1. the two versions (coefficient and FFT embeddings) of ffnp
        are consistent
    2. ffnp output lattice vectors close to the targets.
    """
    f = sign_KAT[n][0]["f"]
    g = sign_KAT[n][0]["g"]
    F = sign_KAT[n][0]["F"]
    G = sign_KAT[n][0]["G"]
    B = [[g, neg(f)], [G, neg(F)]]
    G0 = gram(B)
    G0_fft = [[fft(elt) for elt in row] for row in G0]
    T = ffldl(G0)
    T_fft = ffldl_fft(G0_fft)
    sqgsnorm = gs_norm(f, g, q)
    m = 0
    for i in range(iterations):
        t = [[random() for i in range(n)], [random() for i in
            range(n)]]
        t_fft = [fft(elt) for elt in t]
        z = ffnp(t, T)
        z_fft = ffnp_fft(t_fft, T_fft)
        zb = [ifft(elt) for elt in z_fft]
        zb = [[round(coef) for coef in elt] for elt in zb]
        if z != zb:
            print("ffnp and ffnp_fft are not consistent")
            return False
        diff = [sub(t[0], z[0]), sub(t[1], z[1])]
        diffB = vecmatmul(diff, B)
        norm_zmc = int(round(sqnorm(diffB)))
        m = max(m, norm_zmc)
    th_bound = (n / 4.) * sqgsnorm
    if m > th_bound:
        print("Warning: ffnp does not output vectors as short as
            expected")
        return False
    else:
        print("m={}, th_bound={:.3f}".format(m, th_bound))
        print("ffnp output vectors as short as expected since m
            <= th_bound")
```

```
              return True
if __name__ == "__main__":
    n = 512       # select Falcon-512 or Falcon-1024
    cases = 5     # Number of tests
    print("** Testing ffNP of Falcon-",n)
    for i in range(cases):
        print("\n<< Test Case :", i+1,">>")
        test_ffnp(n,i)
print("\nTest passed:")
```

Script 6.10 Test program for `ffnp()`

```
** Testing ffNP of Falcon-512

<< Test Case : 1 >>
m=0, th_bound=2136281.741
ffnp output vectors as short as expected since m <= th_bound
<< Test Case : 2 >>
m=1020056, th_bound=2136281.741
ffnp output vectors as short as expected since m <= th_bound
<< Test Case : 3 >>
m=1084552, th_bound=2136281.741
ffnp output vectors as short as expected since m <= th_bound
<< Test Case : 4 >>
m=1083516, th_bound=2136281.741
ffnp output vectors as short as expected since m <= th_bound
<< Test Case : 5 >>
m=1108812, th_bound=2136281.741
ffnp output vectors as short as expected since m <= th_bound
Test passed:
```

Fig. 6.12 Output of 5 `ffnp()` tests for FALCON-512

The value of n is fixed at 512 or 1024 depending on which version of FALCON you are testing. Figs. 6.12 and 6.13 show the printout for 5 test cases `ffnp()` function used for FALCON-512 and FALCON-1024, respectively.

6.2.4 Checking `falcon.py`

This section describes the correctness of executing FALCON-512 and FALCON-1024 from the predetermined polynomials, $f, g, F, and\ G$ which is provided as `falcon.py` in the FALCON Python package simply. The test script is listed as Script 6.11. The value of n is fixed at 512 or 1024 depending on which version of FALCON you are verifying.

6.2 FALCON-Specific Modules

```
** Testing ffNP of Falcon-1024

<< Test Case : 1 >>
m=0, th_bound=4228608.000
ffnp output vectors as short as expected since m  <= th_bound
<< Test Case : 2 >>
m=2187656, th_bound=4228608.000
ffnp output vectors as short as expected since m  <= th_bound
<< Test Case : 3 >>
m=2144881, th_bound=4228608.000
ffnp output vectors as short as expected since m  <= th_bound
<< Test Case : 4 >>
m=2139078, th_bound=4228608.000
ffnp output vectors as short as expected since m  <= th_bound
<< Test Case : 5 >>
m=2150386, th_bound=4228608.000
ffnp output vectors as short as expected since m  <= th_bound
Test passed:
```

Fig. 6.13 Output of 5 ffnp() tests for FALCON-1024

```
1   from common import q
2   from falcon import SecretKey, PublicKey
3   from scripts.sign_KAT import sign_KAT
4
5   import random
6   import string
7
8   # Function to generate a random message
9   def generate_random_message(length=26):
10      letters = string.ascii_lowercase  # Lowercase letters a-z
11      message = ''.join(random.choice(letters) for i in
            range(length))
12      return message.encode()  # Encoding the message as bytes
13
14  def test_f_signature(n, iterations=1):
15      f = sign_KAT[n][0]["f"]
16      g = sign_KAT[n][0]["g"]
17      F = sign_KAT[n][0]["F"]
18      G = sign_KAT[n][0]["G"]
19
20      sk = SecretKey(n, [f, g, F, G])
21      print("== Leading 10 values of private key")
22      print("f = ".ljust(3) + "[" + ", ".join(["{}".format(x) for
            x in f[:10]]) + ", ...]")
23      print("g = ".ljust(3) + "[" + ", ".join(["{}".format(x) for
            x in g[:10]]) + ", ...]")
24      print("F = ".ljust(3) + "[" + ", ".join(["{}".format(x) for
            x in F[:10]]) + ", ...]")
25      print("G = ".ljust(3) + "[" + ", ".join(["{}".format(x) for
            x in G[:10]]) + ", ...]")
```

```
    pk = PublicKey(sk)
    print("== Leading 10 values of public key",)
    print("h = ".ljust(3) + "[" + ", ".join(["{}".format(x) for
        x in pk.h[:10]]) + ", ...]")

    for i in range(iterations):
        message = generate_random_message()
        print("Messge  = ",str(message))

        sig = sk.sign(message); sig_str= sig.hex()
        print("Signature =", sig_str[:30], '...', sig_str[-20:])
        print("Length of Signature:",
             int(len(sig_str)/2),"Bytes");

        if (pk.verify(message, sig)== True):
            print("Verification passed!!")
        else:
            print("Verification failed!!")
            return False
    return True
if __name__ == "__main__":
    n = 512    # select Falcon-512 or Falcon-1024
    cases = 3  # Number of tests
    print("** Testing keygen, sign and verify procedures of
         Falcon-",n)
    for i in range(cases):
        print("\n<< Test Case :", i+1,">>")
        test_f_signature(n,i) # degree of cyclotomic poly.
             (power of 2)
print("\nTest passed:")
```

Script 6.11 Test script for `falcon.py`

By setting the value of n at line 46 of Script 6.11 at 512 or 1024, Figs. 6.14 and 6.15 present three examples of built-in key pairs, a random message, its signature in hexadecimal notation, the verification of signature for FALCON-512 and FALCON-1024, respectively.

6.2.5 Checking `test.py`

This section discusses the results of `test.py` provided in the FALCON Python package. Fig. 6.16 shows the specifications of the computer used in executing `test.py`, with the full Python script available in the FALCON Python package.

```
from common import q, sqnorm
from fft import add, sub, mul, div, neg, fft, ifft
from ntt import mul_zq, div_zq
from samplerz import samplerz, MAX_SIGMA
from ffsampling import ffldl, ffldl_fft, ffnp, ffnp_fft
```

6.2 FALCON-Specific Modules

```
** Testing keygen, sign and verify procedures of FALCON-512

<< Test Case : 1 >>
== Leading 10 values of private key
f = [1, -3, 0, 4, 0, 5, -3, -4, 4, -2, ...]
g = [-4, -7, 4, -2, 3, 3, -2, 4, -7, -1, ...]
F = [30, -32, -19, 0, -14, 46, -28, -18, 1, 19, ...]
G = [-25, -14, 10, 8, 28, 18, 7, 12, 34, -18, ...]
== Leading 10 values of public key
h = [11496, 8750, 6367, 8513, 9698, 2801, 11184, 7720, 3044, 6551, ...]

<< Test Case : 2 >>
== Leading 10 values of private key
f = [1, -3, 0, 4, 0, 5, -3, -4, 4, -2, ...]
g = [-4, -7, 4, -2, 3, 3, -2, 4, -7, -1, ...]
F = [30, -32, -19, 0, -14, 46, -28, -18, 1, 19, ...]
G = [-25, -14, 10, 8, 28, 18, 7, 12, 34, -18, ...]
== Leading 10 values of public key
h = [11496, 8750, 6367, 8513, 9698, 2801, 11184, 7720, 3044, 6551, ...]
Messge  = b'zyggautvsowlwphccrpgbaxlcj'
Signature = 393d488cde1b60858f3c5c23944a81 ... 00000000000000000000
Length of Signature: 666 Bytes
Verification passed!!

<< Test Case : 3 >>
== Leading 10 values of private key
f = [1, -3, 0, 4, 0, 5, -3, -4, 4, -2, ...]
g = [-4, -7, 4, -2, 3, 3, -2, 4, -7, -1, ...]
F = [30, -32, -19, 0, -14, 46, -28, -18, 1, 19, ...]
G = [-25, -14, 10, 8, 28, 18, 7, 12, 34, -18, ...]
== Leading 10 values of public key
h = [11496, 8750, 6367, 8513, 9698, 2801, 11184, 7720, 3044, 6551, ...]
Messge  = b'xfhnneoyqmqtndbaizmhujeycn'
Signature = 396d1dee149ee6a37f8e131f76d7a0 ... 00000000000000000000
Length of Signature: 666 Bytes
Verification passed!!
Messge  = b'jrcnohrzzrtysfrbedikqjpqnf'
Signature = 395660877d26f70a301e439f0bc2e8 ... 00000000000000000000
Length of Signature: 666 Bytes
Verification passed!!

Test passed:
```

Fig. 6.14 Three examples of key pairs and signature executing FALCON-512

```
** Testing keygen, sign and verify procedures of FALCON-1024

<< Test Case : 1 >>
== Leading 10 values of private key
f = [2, 1, 3, 3, 2, 2, 2, -4, 2, 0, ...]
g = [3, -3, -3, -1, 2, 3, 1, -2, 0, 2, ...]
F = [14, -13, -33, 46, 31, 8, 12, 29, 22, -2, ...]
G = [-58, 19, -8, 0, 6, -11, 2, 1, -1, 20, ...]
== Leading 10 values of public key
h = [3680, 7862, 6250, 919, 1038, 11753, 2971, 2770, 12273, 2831, ...]

<< Test Case : 2 >>
== Leading 10 values of private key
f = [2, 1, 3, 3, 2, 2, 2, -4, 2, 0, ...]
g = [3, -3, -3, -1, 2, 3, 1, -2, 0, 2, ...]
F = [14, -13, -33, 46, 31, 8, 12, 29, 22, -2, ...]
G = [-58, 19, -8, 0, 6, -11, 2, 1, -1, 20, ...]
== Leading 10 values of public key
h = [3680, 7862, 6250, 919, 1038, 11753, 2971, 2770, 12273, 2831, ...]
Messge  = b'rtkczofilqmoajhnnjrzycsojd'
Signature = 3a0b6d34bfc1b8baf6c08a409f9bbf ... bc24999fe5424e1a0000
Length of Signature: 1280 Bytes
Verification passed!!

<< Test Case : 3 >>
== Leading 10 values of private key
f = [2, 1, 3, 3, 2, 2, 2, -4, 2, 0, ...]
g = [3, -3, -3, -1, 2, 3, 1, -2, 0, 2, ...]
F = [14, -13, -33, 46, 31, 8, 12, 29, 22, -2, ...]
G = [-58, 19, -8, 0, 6, -11, 2, 1, -1, 20, ...]
== Leading 10 values of public key
h = [3680, 7862, 6250, 919, 1038, 11753, 2971, 2770, 12273, 2831, ...]
Messge  = b'cjsxpaqeohtuovrytwlgxzclpc'
Signature = 3a7cea22a29574edb4e11473f9eb3a ... f9b8bcfb254f93000000
Length of Signature: 1280 Bytes
Verification passed!!
Messge  = b'hdytylffiiipohcmxlrvyuyqde'
Signature = 3ababef5736137d8eba24de16e1163 ... 00000000000000000000
Length of Signature: 1280 Bytes
Verification passed!!

Test passed:
```

Fig. 6.15 Three examples of key pairs and signature executing FALCON-1024

```
CPU        Intel(R) Core(TM) i7-8700 CPU @ 3.20GHz   3.19 GHz
RAM        32.0GB
System     64 bit OS, x64
Edition        Windows 10 Pro
Version        22H2
```

Fig. 6.16 Specification of my test computer used in test.py

6.2 FALCON-Specific Modules

```python
from ffsampling import gram
from random import randint, random, gauss, uniform
from math import sqrt, ceil
from ntrugen import karamul, ntru_gen, gs_norm
from falcon import SecretKey, PublicKey, Params
from falcon import SALT_LEN, HEAD_LEN, SHAKE256
from encoding import compress, decompress
from scripts import saga
from scripts.samplerz_KAT512 import sampler_KAT512
from scripts.sign_KAT import sign_KAT
from scripts.samplerz_KAT1024 import sampler_KAT1024
# https://stackoverflow.com/a/25823885/4143624
from timeit import default_timer as timer

def vecmatmul(t, B):
    """Compute the product t * B, where t is a vector and B is a
        square matrix.

    Args:
        B: a matrix

    Format: coefficient
    """
    nrows = len(B)
    ncols = len(B[0])
    deg = len(B[0][0])
    assert(len(t) == nrows)
    v = [[0 for k in range(deg)] for j in range(ncols)]
    for j in range(ncols):
        for i in range(nrows):
            v[j] = add(v[j], mul(t[i], B[i][j]))
    return v

def test_fft(n, iterations=10):
    """Test the FFT."""
    for i in range(iterations):
        f = [randint(-3, 4) for j in range(n)]
        g = [randint(-3, 4) for j in range(n)]
        h = mul(f, g)
        k = div(h, f)
        k = [int(round(elt)) for elt in k]
        if k != g:
            print("(f * g) / f =", k)
            print("g =", g)
            print("mismatch")
            return False
    return True

def test_ntt(n, iterations=10):
    """Test the NTT."""
    for i in range(iterations):
        f = [randint(0, q - 1) for j in range(n)]
        g = [randint(0, q - 1) for j in range(n)]
        h = mul_zq(f, g)
```

```
            try:
                k = div_zq(h, f)
                if k != g:
                    print("(f * g) / f =", k)
                    print("g =", g)
                    print("mismatch")
                    return False
            except ZeroDivisionError:
                continue
    return True


def check_ntru(f, g, F, G):
    """Check that f * G - g * F = q mod (x ** n + 1)."""
    a = karamul(f, G)
    b = karamul(g, F)
    c = [a[i] - b[i] for i in range(len(f))]
    return ((c[0] == q) and all(coef == 0 for coef in c[1:]))


def test_ntrugen(n, iterations=10):
    """Test ntru_gen."""
    for i in range(iterations):
        f, g, F, G = ntru_gen(n)
        if check_ntru(f, g, F, G) is False:
            return False
    return True


def test_ffnp(n, iterations):
    """Test ffnp.

    This functions check that:
    1. the two versions (coefficient and FFT embeddings) of ffnp
       are consistent
    2. ffnp output lattice vectors close to the targets.
    """
    f = sign_KAT[n][0]["f"]
    g = sign_KAT[n][0]["g"]
    F = sign_KAT[n][0]["F"]
    G = sign_KAT[n][0]["G"]
    B = [[g, neg(f)], [G, neg(F)]]
    G0 = gram(B)
    G0_fft = [[fft(elt) for elt in row] for row in G0]
    T = ffldl(G0)
    T_fft = ffldl_fft(G0_fft)
    sqgsnorm = gs_norm(f, g, q)
    m = 0
    for i in range(iterations):
        t = [[random() for i in range(n)], [random() for i in
            range(n)]]
        t_fft = [fft(elt) for elt in t]
        z = ffnp(t, T)
        z_fft = ffnp_fft(t_fft, T_fft)

        zb = [ifft(elt) for elt in z_fft]
```

6.2 FALCON-Specific Modules

```
            zb = [[round(coef) for coef in elt] for elt in zb]
            if z != zb:
                print("ffnp and ffnp_fft are not consistent")
                return False
            diff = [sub(t[0], z[0]), sub(t[1], z[1])]
            diffB = vecmatmul(diff, B)
            norm_zmc = int(round(sqnorm(diffB)))
            m = max(m, norm_zmc)
        th_bound = (n / 4.) * sqgsnorm
        if m > th_bound:
            print("Warning: ffnp does not output vectors as short as
                expected")
            return False
        else:
            return True

def test_compress(n, iterations):
    """Test compression and decompression."""
    try:
        sigma = 1.5 * sqrt(q)
        slen = Params[n]["sig_bytelen"] - SALT_LEN - HEAD_LEN
    except KeyError:
        return True
    for i in range(iterations):
        while(1):
            initial = [int(round(gauss(0, sigma))) for coef in
                range(n)]
            compressed = compress(initial, slen)
            if compressed is not False:
                break
        decompressed = decompress(compressed, slen, n)
        if decompressed != initial:
            return False
    return True

def test_samplerz(nb_mu=100, nb_sig=100, nb_samp=1000):
    """
    Test our Gaussian sampler on a bunch of samples.
    This is done by using a light version of the SAGA test suite,
    see ia.cr/2019/1411.
    """
    # Minimal size of a bucket for the chi-squared test (must be
        >= 5)
    chi2_bucket = 10
    assert(nb_samp >= 10 * chi2_bucket)
    sigmin = 1.3
    nb_rej = 0
    for i in range(nb_mu):
        mu = uniform(0, q)
        for j in range(nb_sig):
            sigma = uniform(sigmin, MAX_SIGMA)
            list_samples = [samplerz(mu, sigma, sigmin) for _ in
                range(nb_samp)]
            v = saga.UnivariateSamples(mu, sigma, list_samples)
```

```
            if (v.is_valid is False):
                nb_rej += 1
    return True
    if (nb_rej > 5 * ceil(saga.pmin * nb_mu * nb_sig)):
        return False
    else:
        return True


def KAT_randbytes(k):
    """
    Use a fixed bytestring 'octets' as a source of random bytes
    """
    global octets
    oc = octets[: (2 * k)]
    if len(oc) != (2 * k):
        raise IndexError("Randomness string out of bounds")
    octets = octets[(2 * k):]
    return bytes.fromhex(oc)[::-1]

def test_samplerz_KAT(unused, unused2):
    # octets is a global variable used as samplerz's randomness.
    # It is set to many fixed values by test_samplerz_KAT,
    # then used as a randomness source via KAT_randbits.
    global octets
    for D in sampler_KAT512 + sampler_KAT1024:
        mu = D["mu"]
        sigma = D["sigma"]
        sigmin = D["sigmin"]
        # Hard copy. octets is the randomness source for samplez
        octets = D["octets"][:]
        exp_z = D["z"]
        try:
            z = samplerz(mu, sigma, sigmin,
                randombytes=KAT_randbytes)
        except IndexError:
            return False
        if (exp_z != z):
            print("SamplerZ does not match KATs")
            return False
    return True

def test_signature(n, iterations=10):
    """
    Test Falcon.
    """
    f = sign_KAT[n][0]["f"]
    g = sign_KAT[n][0]["g"]
    F = sign_KAT[n][0]["F"]
    G = sign_KAT[n][0]["G"]
    sk = SecretKey(n, [f, g, F, G])
    pk = PublicKey(sk)
    for i in range(iterations):
        message = b"abc"
```

6.2 FALCON-Specific Modules

```python
            sig = sk.sign(message)
            if pk.verify(message, sig) is False:
                return False
        return True


def test_sign_KAT():
    """
    Test the signing procedure against test vectors obtained from
    the Round 3 implementation of Falcon.

    Starting from the same private key, same message, and same
        SHAKE256
    context (for randomness generation), we check that we obtain
        the
    same signatures.
    """
    message = b"data1"
    shake = SHAKE256.new(b"external")
    for n in sign_KAT:
        sign_KAT_n = sign_KAT[n]
        for D in sign_KAT_n:
            f = D["f"]
            g = D["g"]
            F = D["F"]
            G = D["G"]
            sk = SecretKey(n, [f, g, F, G])
            # The next line is done to synchronize the SHAKE256
                context
            # with the one in the Round 3 C implementation of
                Falcon.
            _ = shake.read(8 * D["read_bytes"])
            sig = sk.sign(message, shake.read)
            if sig != bytes.fromhex(D["sig"]):
                return False
    return True


def wrapper_test(my_test, name, n, iterations):
    """
    Common wrapper for tests. Run the test, print whether it is
        successful,
    and if it is, print the running time of each execution.
    """
    d = {True: "OK      ", False: "Not OK"}
    start = timer()
    rep = my_test(n, iterations)
    end = timer()
    message = "Test {name}".format(name=name)
    message = message.ljust(20) + ": " + d[rep]
    if rep is True:
        diff = end - start
        msec = round(diff * 1000 / iterations, 3)
        message += " ({msec} msec /
            execution)".format(msec=msec).rjust(30)
    print(message)
```

```python
# Dirty trick to fit test_samplerz into our test wrapper
def test_samplerz_simple(n, iterations):
    return test_samplerz(10, 10, iterations // 100)

def test(n, iterations=500):
    """A battery of tests."""
    wrapper_test(test_fft, "FFT", n, iterations)
    wrapper_test(test_ntt, "NTT", n, iterations)
    # test_ntrugen is super slow, hence performed over a single
        iteration
    wrapper_test(test_ntrugen, "NTRUGen", n, 1)
    wrapper_test(test_ffnp, "ffNP", n, iterations)
    # test_compress and test_signature are only performed
    # for parameter sets that are defined.
    if (n in Params):
        wrapper_test(test_compress, "Compress", n, iterations)
        wrapper_test(test_signature, "Signature", n, iterations)
    #     wrapper_test(test_sign_KAT, "Signature KATs", n,
            iterations)
    print("")
# Run all the tests
if (__name__ == "__main__"):
    print("Test Sig KATs         : ", end="")
    print("OK" if (test_sign_KAT() is True) else "Not OK")
    # wrapper_test(test_samplerz_simple, "SamplerZ", None,
        100000)
    # raise ValueError(msg)
    #ValueError: For each axis slice, the sum of the observed
        frequencies must agree with the sum of the expected
        frequencies to a relative tolerance of 1e-08, but the
        percent differences are:
    # 0.002004008016032064
    wrapper_test(test_samplerz_KAT, "SamplerZ KATs", None, 1)
    print("")

    for i in range(9, 11):
        n = (1 << i)
        it = 1000
        print("Test battery for n = {n}".format(n=n))
        test(n, it)
```

Script 6.12 Test script for falcon.py

Fig. 6.17 shows the output from executing test.py, which includes testing the FALCON Signature Known Answer Test (KAT) and the FALCON SamplerZ KAT. The tests are conducted for rings of degree $n = 512$ and $n = 1024$. Also, Fig. 6.17 displays the average time taken for various operations—Test FFT, Test NTT, Test NTRUGen, Test ffNP, Test Compress, and Test Signature—after iterating 1,000 times. The time consumed may vary slightly depending on your test computer.

6.2 FALCON-Specific Modules

```
Test Sig KATs          : OK
Test SamplerZ KATs     : OK          (37.693 msec / execution)

Test battery for n = 512
Test FFT               : OK          (8.091 msec / execution)
Test NTT               : OK          (8.824 msec / execution)
Test NTRUGen           : OK          (4827.332 msec / execution)
Test ffNP              : OK          (52.428 msec / execution)
Test Compress          : OK          (1.458 msec / execution)
Test Signature         : OK          (29.658 msec / execution)

Test battery for n = 1024
Test FFT               : OK          (17.68 msec / execution)
Test NTT               : OK          (19.158 msec / execution)
Test NTRUGen           : OK          (16984.991 msec / execution)
Test ffNP              : OK          (118.489 msec / execution)
Test Compress          : OK          (2.865 msec / execution)
Test Signature         : OK          (60.529 msec / execution)
```

Fig. 6.17 Time consumed in msec executing test.py

Chapter 7
Checking SOLMAE with Python

SOLMAE Python package is available at web page: https://solmae-sign.info. This repository contains the following files (roughly in order of dependency):

1. common.py contains shared functions and constants
2. encoding.py implements compress and decompress
3. rng.py implements a ChaCha20-based PRNG(standalone)
4. samplerz.py implements a Gaussian sampler over the integers (standalone)
5. fft_constants.py contains precomputed constants used in the FFT
6. ntt_constants.py contains precomputed constants used in the NTT
7. fft.py implements the FFT over $R[x]/(x^n + 1)$
8. ntt.py implements the NTT over $Z_q[x]/(x^n + 1)$
9. ntrugen.py generate polynomials f, g, F, G in $Z[x]/(x^n + 1)$ such that $f \cdot G - g \cdot F = q$
10. params.py contains security parameters
11. Unifcrown.py implements Unifcrown sampler and its test script
12. Pairgen.py implements Pairgen and its test script
13. keygen.py implements keygen and its test script
14. PeikertSampler.py implements Peikert Sampler
15. N_sampler.py implements N-sampler
16. Sampler.py implements Sampler
17. solmae.py implements keygen, sign and verify procedures of SOLMAE-512 or SOLMAE-1024
18. test.py contains how to use and to check that everything is properly implemented.(same as FALCON Python Package)

To implement the SOLMAE in Python, the modules used for FALCON such as common.py, encoding.py, rng.py, samplerz.py, fft_constants.py, fft.py, ntt_constants.py, ntt.py, and ntrugen.py are re-used, as their functionalities are also essential for the operation of SOLMAE.

7.1 SOLMAE-Specific Modules

This section outlines the specific functions and operations exclusive to both SOLMAE-512 and SOLMAE-1024. It includes generating security parameters, generating pairs f and g, creating private and public keys, employing a uniform crown sampler for random number, and Gaussian samplers including all procedure tests of SOLMAE-512 and SOLMAE-1024.

7.1.1 Checking parameters.py

Under the folder ..\..\script, there is a file named solmae_params.py as shown in Script 7.1. This script has been adapted from parameters.py in FALCON Python package to be suitable for SOLMAE.

```
from math import sqrt, exp, log, pi, floor
# Constants
e = exp(1)
q = 1024 * 12 + 1   # Modulo
NB_QUERIES = 2 ** 64   # NIST Recommendation
eta = 1 / (2 ** 41)
sigmax = 1.8205   # Max. acceptable std. dev for Gaussian sampler
DEBUG = True
def smooth(eps, n, normalized=True):
    """Calculate the smoothing parameter."""
    rep = sqrt(log(2 * n * (1 + 1 / eps)) / pi)
    if normalized:
        return rep / sqrt(2 * pi)
    else:
        return rep
def ssmooth(eps, n):
    """Estimation of the smoothing parameter of ZZ^n."""
    return sqrt(log(2 * n * (1 + 1 / eps)) / pi) / sqrt(2 * pi)
def dimensionsforfree(B):
    """Calculate dimensions for free."""
    return round(B * log(4 / 3) / log(B / (2 * pi * exp(1))))
def print_security(B):
    """Print the security parameters."""
    sec_qrec_classical = floor(B * 0.292)
    sec_qrec_quantum = floor(B * 0.265)
    print(f"BIKZ:\t{B}")
    print(f"Classical:\t{sec_qrec_classical}")
    print(f"Quantum:\t{sec_qrec_quantum}")
def compute_para(d, alp, delt, corr):
    """Compute parameters based on input."""
    gs_norm = alp * sqrt(q)
    smoothing = 1 / pi * sqrt(1 / 2 * log(2 * d * (1 + 1 / eta)))
    sigma_sig = smoothing * alp * sqrt(q)
    gamma = corr * sigma_sig * sqrt(2 * d)
```

7.1 SOLMAE-Specific Modules

```python
        R_minus = (1 / alp + delt) * sqrt(q)
        R_plus = (alp - delt) * sqrt(q)
        return gs_norm, smoothing, sigma_sig, gamma, R_minus, R_plus
def solmae_security(n, sigma_offset, fg_norm, target_bitsec,
    target_rejection=0.1, verbose=True):
    """Calculate security parameters for SOLMAE."""
    eps = 1 / sqrt(target_bitsec * NB_QUERIES)
    sigma = sigma_offset * sqrt(q) * smooth(eps, 2 * n)
    tau = 1.1  # Estimate of the signature size w.r.t rejection
        prob.

    while True:
        max_sig_norm = floor(tau * sqrt(2 * n) * sigma)
        rejection_rate = exp(2 * n * (1 - tau ** 2) / 2) * tau
            ** (2 * n)
        if rejection_rate > target_rejection:
            break
        else:
            tau -= 0.001

    B = 100  # Initial block for Key recovery
    sigma_fg = fg_norm / sqrt(2 * n)

    while True:
        left = (B / (2 * pi * e)) ** (1 - n / B) * sqrt(q)
        right = sqrt(3 * B / 4) * sigma_fg
        if left > right:
            break
        else:
            B += 1
    if verbose:
        print("  -----[ Key Recovery ] -----")
        print_security(B)

    B = 100  # Signature forgery

    def condition_LH(beta):
        return min([(((pi * B) ** (1 / B) * B / (2 * pi * e)) **
            ((2 * n - k) / (2 * B - 2))) *
                    q ** (n / (2 * n - k)) for k in range(n)])
    while condition_LH(B) > max_sig_norm:
        B += 1
    sec_forgery_classical = (B * 0.292)
    sec_forgery_quantum = (B * 0.265)
    if verbose:
        print("  -----[ Signature forgery ] -----")
        print_security(B)
    return
def para_gen():
    input_para = [
        {"d": 512, "alpha": 1.17, "delta": 0.065, "correction":
            1.04},
        {"d": 1024, "alpha": 1.64, "delta": 0.3, "correction":
            1.04}
```

```python
    ]
    for params in input_para:
        d = params["d"]
        alp = params["alpha"]
        delt = params["delta"]
        corr = params["correction"]
        gs_norm, smoothing, sigma_sig, gamma, R_minus, R_plus = \
            compute_para(d, alp, delt, corr)

        if DEBUG:
            print(f"\n ** Security parameters for d = {d} **")
            print("alpha(quality)       = \
                {:..3f}".format(alp).ljust(25))
            print("GS_norm              = \
                {:..3f}".format(gs_norm).ljust(25))
            print("smoothing            = \
                {:..3f}".format(smoothing).ljust(25))
            print("sigma_sig(sig width) = \
                {:..3f}".format(sigma_sig).ljust(25))
            print("gamma                = \
                {:..3f}".format(gamma).ljust(25))
            print("gamma^2(sig. bound) = {:..3f}".format(gamma * \
                gamma).ljust(25))
            print("R_minus              = \
                {:..3f}".format(R_minus).ljust(25))
            print("R_plus               = \
                {:..3f}".format(R_plus).ljust(25))
    print("\n== C/Q security of SOLMAE_512 ==")
    solmae_security(512, 2.04, 1.17 * sqrt(q), 128, verbose=True)
    print("\n== C/Q security of SOLMAE_1014 ==")
    solmae_security(1024, 2.33, 1.17 * sqrt(q), 256,
        verbose=True)
if __name__ == "__main__":
    para_gen()
print("Test passed.")
```

Script 7.1 solmae_params.py

Fig. 7.1 displays the security parameters used to configure the programs for SOLMAE-512 and SOLMAE-1024 along with their estimated classical and quantum security levels.

7.1.2 Checking Unifcrown.py

The Python script shown in Script 7.2 is designed to efficiently generate random values in an annular region with a fixed radius. This approach ensures that the values are uniformly distributed within the specified annular region, making it suitable for applications that require random sampling within such geometric constraints.

7.1 SOLMAE-Specific Modules

```
 ** Security parameters for d = 512 **
alpha(quality)      = 1.170
GS_norm             = 129.701
smoothing           = 1.338
sigma_sig(sig width) = 173.571
gamma               = 5776.442
gamma^2(sig. bound) = 33367279.487
R_minus             = 101.954
R_plus              = 122.496

 ** Security parameters for d = 1024 **
alpha(quality)      = 1.640
GS_norm             = 181.803
smoothing           = 1.351
sigma_sig(sig width) = 245.670
gamma               = 11562.453
gamma^2(sig. bound) = 133690311.392
R_minus             = 100.852
R_plus              = 148.547

== C/Q security of SOLMAE_512 ==
 -----[ Key Recovery ] -----
BIKZ:    458
Classical:       133
Quantum:         121
 -----[ Signature forgery ] -----
BIKZ:    349
Classical:       101
Quantum:         92

== C/Q security of SOLMAE_1014 ==
 -----[ Key Recovery ] -----
BIKZ:    936
Classical:       273
Quantum:         248
 -----[ Signature forgery ] -----
BIKZ:    798
Classical:       233
Quantum:         211
Test passed.
```

Fig. 7.1 Output of solmae_params.py

```
1  ###########################
2  # This is for SOLMAE only.
3  ###########################
4  import numpy as np
5  from rng import ChaCha20
6  from os import urandom
7  import matplotlib.pyplot as plt
8  def Unifcrown(R_min, R_max, randombytes=urandom):
```

```
    u_rho = int.from_bytes(randombytes(8), 'little')
    u_theta = int.from_bytes(randombytes(8), 'little')
    u_rho = (u_rho & 0x1fffffffffffff) * 2**(-53)
    u_theta = (u_theta & 0x1fffffffffffff) * 2**(-53)
    rho = np.sqrt(R_min**2+u_rho*(R_max**2-R_min**2))
    x = rho*np.cos(np.pi/2*u_theta)
    y = rho*np.sin(np.pi/2*u_theta) # equivalent to Algorithm 9
    return x, y
if __name__ == '__main__':
    x_list = []
    y_list = []
    for _ in range(5000):
        x, y = Unifcrown(2, 5)
        x_list.append(x)
        y_list.append(y)
    plt.plot(x_list,y_list, 'o', markersize=3)
    plt.show()
```

Script 7.2 Unifcrown.py and its plotting script

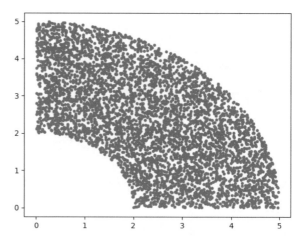

Fig. 7.2 Scatter plot of Unifcrown.py

Fig. 7.2 illustrates a scatter plot of generated by executing Unifcrown.py 5,000 times using matplotlib.pyplot, a widely used tools for creating various types of plots and visualizations. This demonstrates that efficient random generation is effectively achieved.

7.1.3 Checking N_sampler.py

The N_sampler.py script generates random numbers following a Gaussian (or Normal) distribution for the SOLMAE signing procedure, as depicted in Fig. 4.3. To

7.1 SOLMAE-Specific Modules

verify the correctness of the `N_sampler.py` Python script, a test script is provided, shown in Script 7.3. without mentioning `N_sampler.py` Python module.

```python
from rng import ChaCha20
from os import urandom
from params import SOLMAE_D
import numpy as np
import matplotlib.pyplot as plt
import scipy.stats as stats
if __name__ == '__main__':
    # generate datas
    x_list = []; y_list = []; datas = []
    n = 1000 # Set the number of test
    for _ in range(n):
        coeffs = n_sampler(); data = []
        for i in range(SOLMAE_D//2):
            x = coeffs[2*i]; y = coeffs[2*i+1]
            x_list.append(x); y_list.append(y)
            data.append(x);data.append(y)
        datas.append(data)
    # plot the data in 2D
    fig = plt.figure(figsize = (20, 40)); ax1 =
        fig.add_subplot(2, 1, 1)
    plt.title("Distribution represented in 2D")
    plt.plot(x_list,y_list, 'o', markersize=3)
    plt.xlim([-80, 80]); plt.ylim([-80, 80])
    # calculate sample mean and sample variance
    datas = np.array(datas);  mu = np.array([0 for _ in
        range(SOLMAE_D)])
    cov_matrix = SOLMAE_D/2*np.eye(SOLMAE_D)
    cov_matrix_inv = np.linalg.inv(cov_matrix)
    data_mean = np.mean(datas, axis = 0)
    print("mean:", data_mean)
    data_cov_matrix = np.cov(datas.T, ddof = 1)
    print("covariance: ", data_cov_matrix)
    data_cov_matrix_inv = np.linalg.inv(data_cov_matrix)
    # check if the datas follow the distribution N_{d/2} with qq
        plot
    ax2 = fig.add_subplot(2, 1, 2)
    Maha_dist = []
    for data in datas:
        data = np.array(data)
        Maha_dist.append(float(np.dot(np.dot(data.T,
            cov_matrix_inv), data)))
    Maha_dist = np.array(Maha_dist)
    stats.probplot(Maha_dist, dist = stats.chi2(SOLMAE_D),
        plot=ax2)
    plt.title("chi-square QQ-plot")
    plt.show()
```

Script 7.3 Test Python script of `N_sampler.py`

A QQ (Quantile-Quantile) plot is a graphical tool used to compare two probability distributions by plotting their quantiles against each other. It's commonly

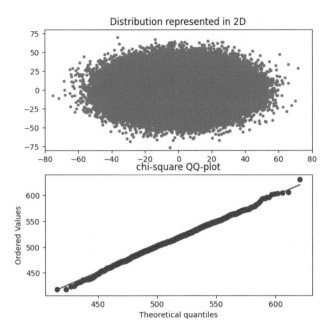

Fig. 7.3 Scatter and QQ plots of checking N_sampler.py

used to assess if a dataset follows a specific theoretical distribution (e.g., normal distribution) provided by `matplotlib.pyplot` Python package.

- Data Quantiles: The quantiles from the sample data are plotted on the y-axis.
- Theoretical Quantiles: The corresponding quantiles from the theoretical distribution (e.g., normal distribution) are plotted on the x-axis.

It the points fall approximately along a straight line (typically the 45° line), the sample data likely follows the theoretical distribution.

Fig. 7.3 shows its Scatter and QQ plots of checking N_sampler.py. The upper part of Fig. 7.3 displays a scatter plot of the data distribution in 2 dimensions, while the lower part shows a chi-square QQ plot. The QQ plot indicates that the generated data follows a Gaussian distribution closely.

7.1.4 Checking Pairgen.py

Verifying the generated short two polynomials f and g used in Algorithm 4, `Pairgen.py` and its test script is shown in Script 7.4.

```
1  ###############################################################
2  #This is (f,g) pair generation and its test program for SOLMAE.
3  ###############################################################
```

7.1 SOLMAE-Specific Modules

```python
import numpy as np
from params import SOLMAE_D, Params, SOLMAE_Q
from Unifcrown import Unifcrown
from os import urandom
from fft import fft , ifft
m_pi = 3.14159265358979323846
count = 0
def Pairgen(randombytes=urandom):
    global count
    R_min = Params[SOLMAE_D]["lower_radius"]
    R_max = Params[SOLMAE_D]["upper_radius"]
    while True:
        flag = True
        count+=1
        f_fft = [0 for _ in range(SOLMAE_D)]
        g_fft = [0 for _ in range(SOLMAE_D)]
        for i in range(SOLMAE_D//2):
            x, y = Unifcrown(R_min, R_max)
            u_1 = int.from_bytes(randombytes(8), 'little')
            theta_x = 2* m_pi *(u_1 & 0x1fffffffffffff) *
                2**(-53)
            u_2 = int.from_bytes(randombytes(8), 'little')
            theta_y = 2* m_pi *(u_2 & 0x1fffffffffffff) *
                2**(-53)
            # multiplied 2pi before this line
            x_re = x*np.cos(theta_x); x_im = x*np.sin(theta_x)
            y_re = y*np.cos(theta_y); y_im = y*np.sin(theta_y)
            f_fft[i] = complex(x_re, x_im)
            f_fft[i + SOLMAE_D//2] = complex(x_re, -x_im)
            g_fft[i] = complex(y_re, y_im)
            g_fft[i+SOLMAE_D//2] = complex(y_re, -y_im)
        f = list(map(lambda n: round(n), ifft(f_fft)))
        g = list(map(lambda n: round(n), ifft(g_fft)))
        res_f_fft = fft(f);  res_g_fft = fft(g)
        for i in range(SOLMAE_D//2):
            norm_sq = res_f_fft[i].real**2 +
                res_f_fft[i].imag**2 +\
                    res_g_fft[i].real**2 + res_g_fft[i].imag**2
            if norm_sq < SOLMAE_Q/Params[SOLMAE_D]["quality"]**2
                or\
                norm_sq >
                    SOLMAE_Q*Params[SOLMAE_D]["quality"]**2:
                # 8977 16822
                flag = False
                continue
        if flag:
            return f, g
if __name__ == '__main__':
    print("==(f,g) pair generation for SOLMAE-", SOLMAE_D)
    cases = 5 # number pf tests
    for i in range(cases):
        count = 0
        f, g = Pairgen()
        print("\n << Test Cases :", i+1, ">>")
```

```
53      print("\count is ", count)
54      print("Leading 10 coeffs of f = ".ljust(8) + "[" +
             ",".join(["{}".format(x) for x in f[:10]]) + ",
             ...]")
55      print("Leading 10 coeffs of g = ".ljust(8) + "[" +
             ",".join(["{}".format(x) for x in g[:10]]) + ",
             ...]")
56   print("Test passed!!")
```

Script 7.4 Pairgen.py and its test script

Fig. 7.4 presents sample output from executing Pairgen.py for SOLMAE-1024 across five cases by setting the value SOLMAE_D = 512 in params.py. Each generation involved rejection sampling, which was performed between 1 (very lucky!) and 106 (very bad!) trials. Despite the variability in the number of sampling attempts, the required short polynomials f and g were successfully generated in every instance.

Similarly, Fig. 7.5 presents sample output from executing Pairgen.py for SOLMAE-1024 across five cases by setting the value SOLMAE_D = 1024 in params.py. Each generation involved rejection sampling, which was performed 1 trial only (very lucky cases). This experiment demonstrates that the required short polynomials f and g were successfully generated in every instance after a single sampling.

```
==(f,g) pair generation for SOLMAE- 512
 << Test Cases : 1 >>
\count is  1
Leading 10 coeffs of f = [-2,2,5,3,0,-2,5,2,1,-2, ...]
Leading 10 coeffs of g = [0,-1,1,5,-1,3,-3,-7,0,-5, ...]
 << Test Cases : 2 >>
\count is  41
Leading 10 coeffs of f = [-2,-2,1,0,0,-1,-3,3,5,9, ...]
Leading 10 coeffs of g = [3,-3,4,-2,-1,1,2,3,7,-4, ...]
 << Test Cases : 3 >>
\count is  106
Leading 10 coeffs of f = [1,1,0,-5,3,-1,0,-3,1,4, ...]
Leading 10 coeffs of g = [1,-3,-5,5,-1,1,1,-1,5,0, ...]
 << Test Cases : 4 >>
\count is  2
Leading 10 coeffs of f = [3,-9,0,1,5,-3,8,1,3,0, ...]
Leading 10 coeffs of g = [7,0,-5,3,2,0,3,4,-4,0, ...]
 << Test Cases : 5 >>
\count is  30
Leading 10 coeffs of f = [-2,-4,-3,4,-7,-2,0,1,-3,-2, ...]
Leading 10 coeffs of g = [-5,4,5,4,-3,1,-4,2,-2,3, ...]
Test passed!!
```

Fig. 7.4 Sample output from executing Pairgen.py for SOLMAE-512

7.1 SOLMAE-Specific Modules

```
==(f,g) pair generation for SOLMAE- 1024
  << Test Cases : 1 >>
count is  1
Leading 10 coeffs of f = [-3,-4,10,2,2,-3,3,2,6,-2, ...]
Leading 10 coeffs of g = [-1,3,-3,2,0,-4,-6,1,2,1, ...]
  << Test Cases : 2 >>
count is  1
Leading 10 coeffs of f = [-4,-1,-1,3,-2,5,-6,2,0,-3, ...]
Leading 10 coeffs of g = [-1,-3,-3,3,-2,-1,-6,0,2,0, ...]
  << Test Cases : 3 >>
count is  1
Leading 10 coeffs of f = [0,5,3,0,-2,0,-3,1,-1,3, ...]
Leading 10 coeffs of g = [0,1,0,3,-1,-4,3,2,-1,-1, ...]
  << Test Cases : 4 >>
count is  1
Leading 10 coeffs of f = [-4,-1,4,-2,1,3,1,-7,-1,2, ...]
Leading 10 coeffs of g = [5,4,-3,2,4,-2,0,-1,2,6, ...]
  << Test Cases : 5 >>
count is  1
Leading 10 coeffs of f = [-1,-3,3,3,-2,1,0,1,-1,2, ...]
Leading 10 coeffs of g = [-2,-1,-5,-3,-4,3,0,4,2,0, ...]
Test passed!!
```

Fig. 7.5 Sample output from executing `Pairgen.py` for SOLMAE-1024

7.1.5 Checking keygen.py

This section describes an example of executing `keygen.py`, which is used for private and public keys of SOLMAE. The details of this example are illustrated in Script 7.5.

```
1  ###############################################################
2  #This is to generate all keys and its test program for SOLMAE.
3  ###############################################################
4  from Pairgen import Pairgen
5  from ntt import ntt, intt, div_ntt
6  from ntrugen import ntru_solve
7  from fft import fft, ifft, add_fft, sub_fft, mul_fft, div_fft,
       adj_fft, cut_half_fft
8  from numpy import sqrt
9  from params import Params, SOLMAE_D
10 from os import urandom
11 class secret_key:
12     def __init__(self):
13         self.f = []; self.g = []; self.F = []; self.G = []
14         self.Sigma1 = []; self.Sigma2 = []
15         self.b10_fft = []; self.b11_fft = []; self.b20_fft = [];
               self.b21_fft = []
16         self.beta10_fft = []; self.beta11_fft = []
17         self.beta20_fft = []; self.beta21_fft = []
```

```python
class public_key:
    def __init__(self):
        self.h = []
def keygen(randombytes=urandom):
    sk = secret_key()
    pk = public_key()
    while True:
        f, g = Pairgen(randombytes)
        try:
            f_ntt = ntt(f)
            g_ntt = ntt(g)
            h_ntt = div_ntt(g_ntt, f_ntt)
        except ZeroDivisionError:
            continue
        try:
            F, G = ntru_solve(f, g)
        except ValueError:
            continue

        sk.f = f; sk.g = g; sk.F = F; sk.G = G
        pk.h = intt(h_ntt)
        break
    # Consistency values (2023.8.10, kkj)
    eta_sq = Params[SOLMAE_D]["smoothing"] ** 2
    sig_width = Params[SOLMAE_D]["signature_width"] ** 2
    eta_sq_fft = [eta_sq for _ in range(SOLMAE_D)]
    sig_width_fft = [sig_width for _ in range(SOLMAE_D)]

    sk.b10_fft = fft(f); sk.b11_fft = fft(g)

    b1_norm = add_fft(mul_fft(adj_fft(sk.b10_fft), sk.b10_fft),
        mul_fft(adj_fft(sk.b11_fft), sk.b11_fft))
    sk.beta10_fft = div_fft(sk.b10_fft, b1_norm)
    sk.beta11_fft = div_fft(sk.b11_fft, b1_norm)
    sk.Sigma1 = [sqrt(elem) for elem in
        cut_half_fft(sub_fft(div_fft(sig_width_fft, b1_norm),
        eta_sq_fft))]
    sk.b20_fft = fft(F); sk.b21_fft = fft(G)
    temp_fft = add_fft(mul_fft(adj_fft(sk.beta10_fft),
        sk.b20_fft), mul_fft(adj_fft(sk.beta11_fft), sk.b21_fft))
    sk.b20_tild_fft = sub_fft(sk.b20_fft, mul_fft(temp_fft,
        sk.b10_fft))
    sk.b21_tild_fft = sub_fft(sk.b21_fft, mul_fft(temp_fft,
        sk.b11_fft))
    b2_tild_norm = add_fft(mul_fft(adj_fft(sk.b20_tild_fft),
        sk.b20_tild_fft), mul_fft(adj_fft(sk.b21_tild_fft),
        sk.b21_tild_fft))
    sk.beta21_fft = div_fft(sk.b21_tild_fft, b2_tild_norm)
    sk.Sigma2 = [sqrt(elem) for elem in
        cut_half_fft(sub_fft(div_fft(sig_width_fft,
        b2_tild_norm), eta_sq_fft))]
    return sk, pk

if __name__ == '__main__':
```

7.1 SOLMAE-Specific Modules

```
sk, pk = keygen()
print("== Leading 3 to 10 values of key pairs for
    SOLMAE_",SOLMAE_D,"==")
print("f              = ".ljust(8) + "[" + ",
    ".join(["{}".format(x) for x in sk.f[:10]]) + ", ...]")
print("g              = ".ljust(8) + "[" + ",
    ".join(["{}".format(x) for x in sk.g[:10]]) + ", ...]")
print("F              = ".ljust(8) + "[" + ",
    ".join(["{}".format(x) for x in sk.F[:10]]) + ", ...]")
print("G              = ".ljust(8) + "[" + ",
    ".join(["{}".format(x) for x in sk.G[:10]]) + ", ...]")
print("h              = ".ljust(8) + "[" + ",
    ".join(["{}".format(x) for x in pk.h[:9]]) + ", ...]")
print("f_fft          = ".ljust(8) + "[" + ",
    ".join(["{:.3f}".format(x) for x in sk.b10_fft[:3]]) +
    ", ...]")
print("g_fft          = ".ljust(8) + "[" + ",
    ".join(["{:.3f}".format(x) for x in sk.b11_fft[:3]]) +
    ", ...]")
print("F_fft          = ".ljust(8) + "[" + ",
    ".join(["{:.3f}".format(x) for x in sk.b20_fft[:3]]) +
    ", ...]")
print("G_fft          = ".ljust(8) + "[" + ",
    ".join(["{:.3f}".format(x) for x in sk.b21_fft[:3]]) +
    ", ...]")
print("beta10_fft     = ".ljust(8) + "[" + ",
    ".join(["{:.3f}".format(x) for x in sk.b10_fft[:3]]) +
    ", ...]")
print("beta11_fft     = ".ljust(8) + "[" + ",
    ".join(["{:.3f}".format(x) for x in sk.b11_fft[:3]]) +
    ", ...]")
print("beta20_fft     = ".ljust(8) + "[" + ",
    ".join(["{:.3f}".format(x) for x in sk.b20_fft[:3]]) +
    ", ...]")
print("beta21_fft     = ".ljust(8) + "[" + ",
    ".join(["{:.3f}".format(x) for x in sk.b21_fft[:3]]) +
    ", ...]")
print("Sigma1         = ".ljust(8) + "[" + ",
    ".join(["{:.3f}".format(x) for x in sk.Sigma1[:3]]) + ",
    ...]")
print("Sigma2         = ".ljust(8) + "[" + ",
    ".join(["{:.3f}".format(x) for x in sk.Sigma2[:3]]) + ",
    ...]")
```

Script 7.5 keygen.py and its test script

Fig. 7.6 shows sample output from executing keygen.py for SOLMAE-512 by setting the value of SOLMAE_D = 512 in params.py including a set of typical values of f, g, F, G, h, f_fft, g_fft, F_fft, G_fft, beta10_fft, beta11_fft, beta20_fft, beta21_fft, Sigma1, and Sigma2.

Similarly, Fig. 7.7 shows sample output from executing keygen.py for SOLMAE-1024 by setting the value of SOLMAE_D = 1024 in params.py including a set of typical values of f, g, F, G, h, f_fft, g_fft, F_fft,

```
== Leading 3 to 10 values of key pairs for SOLMAE-512 ==
f           = [-3, -4, 5, -3, 1, -5, -3, 5, -5, 8, ...]
g           = [-2, -8, -2, -3, 2, 1, 1, -4, -4, 1, ...]
F           = [-15, 26, 22, -21, 4, -29, 1, -31, -34, 26, ...]
G           = [-8, 9, 10, 9, -16, -23, -7, 4, -15, -5, ...]
h           = [1720,8311,1095,2029,11095,5121,10443, 408,11299, ...]
f_fft       = [40.841-14.725j, -43.531-2.073j, -26.580+89.053j, ...]
g_fft       = [94.611-64.152j, 110.894+28.724j, 13.925-43.615j, ...]
F_fft       = [-453.534+135.390j,-233.101+21.042j,219.928-492.749j, ...]
G_fft       = [-802.666+736.641j,329.425+84.522j, -150.752+113.944j,...]
beta10_fft  = [40.841-14.725j,-43.531-2.073j,-26.580+89.053j,..]
beta11_fft  = [94.611-64.152j,110.894+28.724j,13.925-43.615j, ..]
beta20_fft  = [-453.534+135.390j,-233.101+21.042j, 219.928-492.749j,...]
beta21_fft  = [-802.666+736.641j,329.425+84.522j,-150.752+113.944j,...]
Sigma1      = [0.473, 0.463, 1.008, ...]
Sigma2      = [1.092, 1.098, 0.592, ...]
```

Fig. 7.6 Sample output by executing keygen.py for SOLMAE-512

```
== Leading 3 to 10 values of key pairs for SOLMAE-1024 ==
f           = [-2, 3, 2, 5, 2, 6, 1, 6, -1, 3, ...]
g           = [-6, -4, 2, -1, 1, 0, 1, 1, 1, 0, ...]
F           = [-19, 6, -46, 0, -15, -28, 9, -55, 2, -13, ...]
G           = [8, -10, 14, -46, 17, -12, -38, -29, 19, 14, ...]
h           = [6600,2287,11751,416,8459,10370,1998,5697,9809, ...]
f_fft       = [-38.671+13.844j, 29.056+3.182j, -59.014-100.694j, ...]
g_fft       = [-96.583+20.849j, 114.020+3.343j, 11.414+40.975j, ...]
F_fft       = [-128.584-209.765j,-245.246-421.722j,-96.893+1515.268j,...]
G_fft       = [-522.335-641.578j,-667.241-1610.014j,122.029-434.023j,...]
beta10_fft  = [-38.671+13.844j, 29.056+3.182j, -59.014-100.694j,  ...]
beta11_fft  = [-96.583+20.849j, 114.020+3.343j, 11.414+40.975j, ...]
beta20_fft  = [-128.584-209.765j,-245.246-421.722j,-96.893+1515.268j,...]
beta21_fft  = [-522.335-641.578j,-667.241-1610.014j,122.029-434.023j,...]
Sigma1      = [1.856, 1.589, 1.444, ...]
Sigma2      = [1.658, 1.927, 2.083, ...]
```

Fig. 7.7 Sample output by executing keygen.py for SOLMAE-1024

G_fft, beta10_fft, beta11_fft, beta20_fft, beta21_fft, Sigma1, and Sigma2.

7.1.6 Checking solmae.py

This section describes the keygen, sign, and verification procedure of SOLMAE-512 and SOLMAE-1024 from the randomly generated private and its corresponding public key.

7.1 SOLMAE-Specific Modules

The test script is listed as Script 7.6. The value of SOLMAE_D in params.py is fixed at 512 or 1024 depending on which type of SOLMAE you are verifying.

```python
###############################################################
# This is test of keygen, sign and verify procedures of SOLMAE
###############################################################
from solmae import sign, verify
from keygen import secret_key, public_key, keygen
from os import urandom
from params import SOLMAE_D
def test_s_signature(iterations=1):
    sk=secret_key()
    pk=public_key()
    sk, pk = keygen()
    print("==Leading 10 values of keygen function for 
        SOLMAE-",SOLMAE_D)
    print("f         = ".ljust(8) + "[" + ",
        ".join(["{}".format(x) for x in sk.f[:5]]) + ", ...]")
    print("g         = ".ljust(8) + "[" + ",
        ".join(["{}".format(x) for x in sk.g[:5]]) + ", ...]")
    print("F         = ".ljust(8) + "[" + ",
        ".join(["{}".format(x) for x in sk.F[:5]]) + ", ...]")
    print("G         = ".ljust(8) + "[" + ",
        ".join(["{}".format(x) for x in sk.G[:5]]) + ", ...]")
    print("h         = ".ljust(8) + "[" + ",
        ".join(["{}".format(x) for x in pk.h[:5]]) + ", ...]")
    print("f_fft     = ".ljust(8) + "[" + ",
        ".join(["{:.3f}".format(x) for x in sk.b10_fft[:3]]) +
        ", ...]")
    print("g_fft     = ".ljust(8) + "[" + ",
        ".join(["{:.3f}".format(x) for x in sk.b11_fft[:3]]) +
        ", ...]")
    print("F_fft     = ".ljust(8) + "[" + ",
        ".join(["{:.3f}".format(x) for x in sk.b20_fft[:3]]) +
        ", ...]")
    print("G_fft     = ".ljust(8) + "[" + ",
        ".join(["{:.3f}".format(x) for x in sk.b21_fft[:3]]) +
        ", ...]")
    print("beta10_fft = ".ljust(8) + "[" + ",
        ".join(["{:.3f}".format(x) for x in sk.b10_fft[:3]]) +
        ",   ...]")
    print("beta11_fft = ".ljust(8) + "[" + ",
        ".join(["{:.3f}".format(x) for x in sk.b11_fft[:3]]) +
        ",   ...]")
    print("beta20_fft = ".ljust(8) + "[" + ",
        ".join(["{:.3f}".format(x) for x in sk.b20_fft[:3]]) +
        ",   ...]")
    print("beta21_fft = ".ljust(8) + "[" + ",
        ".join(["{:.3f}".format(x) for x in sk.b21_fft[:3]]) +
        ",   ...]")
    print("Sigma1    = ".ljust(8) + "[" + ",
        ".join(["{:.3f}".format(x) for x in sk.Sigma1[:3]]) + ",
        ...]")
```

```python
    print("Sigma2        = ".ljust(8) + "[" + ", 
        ".join(["{:.3f}".format(x) for x in sk.Sigma2[:3]]) + ", 
        ...]")
    message=urandom(SOLMAE_D) #generate SOLMAE-D Bytes message
        randomly
    hex_message = message.hex()
    print("Message =",hex_message[:40], '...', hex_message[-17:])
    print("Length of Message: ", len(message))
    signature = sign(sk, message) # signing
    str_signature =signature.hex()
    print("Signature =",str_signature[:40], '...', 
        str_signature[-17:])
    print("Length of Sig. = ", len(signature))
    if (verify(pk, message, signature)== True): # verifying
        print("Verification passed!!")
    else:
        print("Verification failed!!")
        return False
    return True
if __name__ == "__main__":
    cases = 2 # Number of tests
    print("**Testing of keygen, sign and verify procedures of 
        SOLMAE-",SOLMAE_D)
    for i in range(cases):
        print("\n<< Test Case :", i+1,">>")
        test_s_signature(i) # degree of cyclotomic poly. (power
            of 2)
print("\nTest passed:")
```

Script 7.6 Testing script of solmae.py

Depending on the value of SOLMAE_D in params.py, Figs. 7.8 and 7.9 present two tests of randomly generated key data, a random 512 byte message, its signature in hexadecimal notation, the verification of signature for SOLMAE-512 and SOLMAE-1024, respectively.

7.1 SOLMAE-Specific Modules

```
**Testing of keygen, sign and verify procedures of SOLMAE-512
<< Test Case : 1 >>
==Leading 10 values of keygen function for SOLMAE-512
f           = [-5, 5, 3, 0, 2, ...]
g           = [-2, -1, 0, -5, -3, ...]
F           = [15, -3, 17, -36, -9, ...]
G           = [-29, -4, 22, -4, -37, ...]
h           = [10584, 7983, 3214, 11619, 2601, ...]
f_fft       = [18.000-106.311j, 51.018-54.726j, 98.702-53.016j, ...]
g_fft       = [33.266+53.408j, -56.722+65.946j, -0.474+5.675j, ...]
F_fft       = [491.941+455.316j, 251.152+124.845j, 239.141+322.386j, ...]
G_fft       = [-372.041+103.780j, -185.559-13.212j, 76.266+53.164j, ...]
beta10_fft  = [18.000-106.311j, 51.018-54.726j, 98.702-53.016j, ...]
beta11_fft  = [33.266+53.408j, -56.722+65.946j, -0.474+5.675j, ...]
beta20_fft  = [491.941+455.316j, 251.152+124.845j, 239.141+322.386j, ...]
beta21_fft  = [-372.041+103.780j, -185.559-13.212j, 76.266+53.164j, ...]
Sigma1      = [0.377, 0.705, 0.776, ...]
Sigma2      = [1.148, 0.914, 0.848, ...]
Message = 3ec29088765e80f921aeb648ea26a2f990774b0f ... c24af424a492e304a
Length of Message:  512
Signature = 3904988187c26069f7c51cca41f625f9f53096c2 ... 5b880000000000000
Length of Sig. =  666
Verification passed!!
<< Test Case : 2 >>
==Leading 10 values of keygen function for SOLMAE-512
f           = [0, 1, -5, 0, -3, ...]
g           = [-6, -1, -3, -1, -2, ...]
F           = [51, 4, 2, 22, 51, ...]
G           = [16, -6, 13, 5, 12, ...]
h           = [6845, 7786, 11420, 8450, 5856, ...]
f_fft       = [-61.847+80.484j, 19.041+50.656j, 0.231-62.664j, ...]
g_fft       = [-20.460+49.238j, 99.208-36.054j, -67.567-74.826j, ...]
F_fft       = [176.840+328.914j, -144.564+256.211j, -32.664+83.685j, ...]
G_fft       = [60.618+46.905j, 576.527+74.884j, 52.452+331.064j, ...]
beta10_fft  = [-61.847+80.484j, 19.041+50.656j, 0.231-62.664j, ...]
beta11_fft  = [-20.460+49.238j, 99.208-36.054j, -67.567-74.826j, ...]
beta20_fft  = [176.840+328.914j, -144.564+256.211j, -32.664+83.685j, ...]
beta21_fft  = [60.618+46.905j, 576.527+74.884j, 52.452+331.064j, ...]
Sigma1      = [0.708, 0.592, 0.589, ...]
Sigma2      = [0.912, 1.008, 1.010, ...]
Message = 4334893654155d5775a36745639d558ce560f90e ... 725c8b75fb1be718a
Length of Message:  512
Signature = 39f0ec9b78f3115807b58bb7c4ae478234b8a42e ... 0000000000000000
Length of Sig. =  666
Verification passed!!

Test passed:
```

Fig. 7.8 Two tests of keygen, sign and verify procedures of SOLMAE-512

```
**Testing of keygen, sign and verify procedures of SOLMAE-1024
<< Test Case : 1 >>
==Leading 10 values of keygen function for SOLMAE-1024
f          = [-3, -2, -4, -1, 5, ...]
g          = [2, -1, -2, 1, -2, ...]
F          = [-22, 10, -21, -9, 3, ...]
G          = [30, 3, 58, -15, -14, ...]
h          = [6723, 11180, 8130, 1914, 7378, ...]
f_fft      = [-104.280-96.797j, -69.196-56.365j, 34.935-44.066j, ...]
g_fft      = [21.710+25.918j, 67.878+30.724j, -55.365+120.224j, ...]
F_fft      = [-226.288+2151.829j, 35.686+60.839j, -400.363+278.509j, ...]
G_fft      = [53.959-441.838j, -148.546+45.477j, 896.346-688.553j, ...]
beta10_fft = [-104.280-96.797j, -69.196-56.365j, 34.935-44.066j, ...]
beta11_fft = [21.710+25.918j, 67.878+30.724j, -55.365+120.224j, ...]
beta20_fft = [-226.288+2151.829j, 35.686+60.839j, -400.363+278.509j, ...]
beta21_fft = [53.959-441.838j, -148.546+45.477j, 896.346-688.553j, ...]
Sigma1     = [0.998, 1.624, 1.045, ...]
Sigma2     = [2.592, 1.891, 2.537, ...]
Message = 9ba020beb78f34e3b1f52596c519c754cd6079ed ... d8ba0867d484b8fc6
Length of Message:  1024
Signature = 3adeee535b7f475a7cdcf9fb6ac0874885711af9 ... 0000000000000000
Length of Sig. = 1375
Verification passed!!
<< Test Case : 2 >>
==Leading 10 values of keygen function for SOLMAE-1024
f          = [2, 1, 0, -2, -2, ...]
g          = [1, 0, 2, -1, 0, ...]
F          = [-11, -24, -25, 40, -8, ...]
G          = [24, -24, 6, -12, 13, ...]
h          = [7104, 9935, 3501, 9095, 4081, ...]
f_fft      = [-37.253+120.370j, -33.444-47.082j, 73.358-60.210j, ...]
g_fft      = [82.878-10.505j, 47.505+103.520j, -38.000+51.986j, ...]
F_fft      = [-153.026-842.464j, 136.612-8.505j, -325.009+21.104j, ...]
G_fft      = [-495.479+230.134j, -391.067+139.759j, 310.058+13.235j, ...]
beta10_fft = [-37.253+120.370j, -33.444-47.082j, 73.358-60.210j, ...]
beta11_fft = [82.878-10.505j, 47.505+103.520j, -38.000+51.986j, ...]
beta20_fft = [-153.026-842.464j, 136.612-8.505j, -325.009+21.104j, ...]
beta21_fft = [-495.479+230.134j, -391.067+139.759j, 310.058+13.235j, ...]
Sigma1     = [0.902, 1.369, 1.662, ...]
Sigma2     = [2.703, 2.166, 1.852, ...]
Message = 264c693071384b6f504be07550470cbc07aa6b0e ... a1f380fcbb19e03fa
Length of Message:  1024
Signature = 3a710d9ddb76fd70aac9dc1f4a89d79b6b5ce7f9 ... 0000000000000000
Length of Sig. = 1375
Verification passed!!

Test passed:
```

Fig. 7.9 Two tests of keygen, sign and verify procedures of SOLMAE-1024

Chapter 8
Concluding Remarks

In this monograph, we encapsulate key insights into the comparative analysis between FALCON and SOLMAE, particularly within the evolving landscape of PQC. It underscores the growing necessity for secure algorithms resistant to quantum-based attacks, given the rapid advancements in quantum computing. While FALCON, selected as one of NIST's PQC standard post-quantum signatures, has been widely recognized as a pioneering signature scheme, SOLMAE offers practical advantages such as simpler implementation and enhanced performance.

Both FALCON and SOLMAE ensure message integrity against modification or forgery by attackers, guaranteeing long-term security, even in the future quantum computing era. These schemes require a deeper understanding of algebra compared to DSA or ECDSA, which are based on number theory. To aid in understanding the detailed internal operations of FALCON and SOLMAE, main modules implemented in Python scripts are tested step-by-step, enhancing comprehension for readers with undergraduate-level knowledge.

The key generation, signing, and verification processes of FALCON and SOLMAE depend on a set of security parameters, generated and verified in the `parameters.py` module. The results of these parameters allow configurations for FALCON-512 and SOLMAE-512 (NIST level of security I), as well as FALCON-1024 and SOLMAE-1024 (NIST level of security V). Visualizations of Gaussian and uniform distributions of randomly generated numbers are also provided using Python's visualization and statistical measurement packages.

One of the main concerns in PQC remains side-channel attacks, which are evolving alongside cryptographic algorithms. SOLMAE, by simplifying sampling procedures and introducing parallelizable features, appears better equipped to address these challenges with fewer complexities in masking techniques compared to FALCON. Continuous improvements in performance will be crucial to stay ahead of quantum threats. Moreover, ensuring security across different platform—particularly against power, timing, and EM attacks, etc.—remains a critical issue.

These considerations emphasize the need for ongoing research and optimization in developing robust PQC.

Lastly, as research in this field continues, the cryptographic community must remain vigilant in addressing potential weaknesses, building systems that are not only quantum-resistant but also adaptable to the dynamic needs of an interconnected world.

In Kim's work [21], a performance comparison is presented between FALCON and SOLMAE based on their Python implementations. The study evaluates their efficiency by analyzing execution speed and overall performance to gain a comprehensive understanding. His companion paper in [22] has presented an asymptotic complexity and performance comparison between FALCON and SOLMAE using their C Implementation for a more realistic performance evaluation.

Finally, SOLMAE was established as a Korean TTA standard on Dec. 6, 2024, under the title Quantum–safe Digital Signature based on NTRU Lattices – Part 1: General in Korean, TTAK.KO-12.0410-Part 1 [23] and its Part 2 is planned to be established as a TTA standard in 2025.

References

1. Diffie, W., Hellman, M.E.: New directions in cryptography. IEEE Trans. Inf. Theory **22**(6), 644–654 (1976). https://doi.org/10.1109/TIT.1976.1055638
2. Ducas, L.: Shortest vector from lattice sieving: a few dimensions for free. In: Nielsen, J.B., Rijmen, V. (eds.) EUROCRYPT 2018, Part I, vol. 10820, pp. 125–145. Springer, Heidelberg (2018)
3. Ducas, L., Prest, T.: Fast Fourier orthogonalization. In: Abramov, S.A., Zima, E.V., Gao, X.S. (eds.) Proceedings of the ACM on International Symposium on Symbolic and Algebraic Computation, ISSAC 2016, pp. 191–198. ACM, Waterloo (2016)
4. Ducas, L., Lyubashevsky, V., Prest, T.: Efficient identity-based encryption over NTRU lattices. In: Sarkar, P., Iwata, T. (eds.) ASIACRYPT 2014, Part II. LNCS, vol. 8874, pp. 22–41. Springer, Heidelberg (2014)
5. ElGamal, T.: A public key cryptosystem and a signature scheme based on discrete logarithms. IEEE Trans. Inf. Theory **31**(4), 469–472 (1985). https://doi.org/10.1109/TIT.1985.1057074
6. Espitau, T., Fouque, P.A., Gérard, F., Rossi, M., Takahashi, A., Tibouchi, M., Wallet, A., Yu, Y.: Mitaka: a simpler, parallelizable, maskable variant of FALCON. In: Dunkelman, O., Dziembowski, S. (eds.) Advances in Cryptology – EUROCRYPT 2022, pp. 222–253. Springer International Publishing, Cham (2022)
7. Espitau, T., Fouque, P.A., Gérard, F., Rossi, M., Takahashi, A., Tibouchi, M., Wallet, A., Yu, Y.: Mitaka: a simpler, parallelizable, maskable variant of Falcon. In: Advances in Cryptology, Proc. of EUROCRYPTO 2022, Part III, pp. 222–253 (2022)
8. Espitau, T., Nguyen, T.T.Q., Sun, C., Tibouchi, M., Wallet, A.: Antrag: annular NTRU trapdoor generation. In: Proc. of Asiacrypt2023, Part VII, Guangzhou, pp. 3–32 (2023)
9. Fouque, P.A., Hoffstein, J., Kirchner, P., Lyubashevsky, V., Pornin, T., Prest, T., Ricosset, T., Seiler, G., Whyte, W., Zhang, Z.: Falcon: Fast-fourier lattice-based compact signatures over NTRU. https://falcon-sign.info/
10. Fouque, P.A., Kirchner, P., Tibouchi, M., Wallet, A., Yu, Y.: Key recovery from Gram-Schmidt norm leakage in hash-and-sign signatures over NTRU lattices. Cryptology ePrint Archive, Paper 2019/1180 (2019). https://eprint.iacr.org/2019/1180
11. Gentry, C., Peikert, C., Vaikuntanathan, V.: How to use a short basis: trapdoors for hard lattices and new cryptographic constructions. In: Proc. of 40th ACM STOC 2008, pp. 197–206 (2008)
12. Gentry, C., Peikert, C., Vaikuntanathan, V.: Trapdoors for hard lattices and new cryptographic constructions. In: Proceedings of the 40th Annual ACM Symposium on Theory of Computing (STOC), pp. 197–206. ACM, Victoria (2008). https://doi.org/10.1145/1374376.1374407
13. Goldreich, O., Goldwasser, S., Halevi, S.: Public-key cryptosystems from lattice reduction problems. In: Advances in Cryptology, Proc. of Crypto 1997, pp. 112–131 (1997)

14. Goldwasser, S., Bellare, M.: Lecture Notes on Cryptography. Massachusetts Institute of Technology (2001). Available online at https://cseweb.ucsd.edu/~mihir/papers/gb.html
15. Hoffstein, J., Pipher, J., Silverman, J.H.: NTRU: a ring-based public key cryptosystem. In: Algorithmic Number Theory, Third International Symposium, ANTS-III, Portland, June 21–25, 1998. Lecture Notes in Computer Science, vol. 1423, pp. 267–288. Springer, Berlin (1998)
16. Hoffstein, J., Howgrave-Graham, N., Pipher, J., Silverman, J.H., Whyte, W.: NTRUSIGN: digital signatures using the NTRU lattice. In: Joye, M. (ed.) CT-RSA 2003. LNCS, vol. 2612, pp. 122–140. Springer, Heidelberg (2003)
17. Howe, J., Prest, T., Ricosset, T., Rossi, M.: Isochronous gaussian sampling: from inception to implementation. Cryptology ePrint Archive, Paper 2019/1411 (2019). https://eprint.iacr.org/2019/1411
18. Hulsing, A., Bernstein, D.J., Dobraunig, C., Eichlseder, M., Fluhrer, S., Gazdag, S.L., Kampanakis, P., Kolbl, S., Lange, T., Lauridsen, M.M., Mendel, F., Niederhagen, R., Rechberger, C., Rijneveld, J., Schwabe, P., Aumasson, J.P., Westerbaan, B., Beullens, W.: Sphincs+. https://sphincs.org/
19. IBM: Expanding the IBM quantum roadmap to anticipate the future of quantum-centric supercomputing (2022). https://research.ibm.com/blog/ibm-quantum-roadmap-2025
20. Kahn, D.: The Codebreakers: The Comprehensive History of Secret Communication from Ancient Times to the Internet. Scribner, New York (1996)
21. Kim, K.: Theoretical and empirical analysis of FALCON and SOLMAE using their python implementation. In: Seo, H., Kim, S. (eds.) Information Security and Cryptology – ICISC 2023, pp. 235–260. Springer Nature Singapore, Singapore (2024)
22. Kim, K., Kim, Y.: Asymptotic complexity and performance comparison of FALCON and SOLMAE using their c implementation. Springer Briefs in Information Security and Cryptography, ISBN 978-3-031-81249-1, Springer
23. Kim, K., Kim, Y.: Quantum-safe digital signature based on ntru lattices - part 1: General(in korean). TTAK.KO-12.0410-Part1 (2024)
24. Kim, K., Tibouchi, M., Espitau, T., Takashima, A., Wallet, A., Yu, Y., Guilley, S., Kim, S.: Solmae: algorithm specification. Updated SOLMAE, IRCS Blog (2023). https://ircs.re.kr/?p=1714
25. Koblitz, N.: Elliptic curve cryptosystems. Math. Comput. **48**(177), 203–209 (1987). https://doi.org/10.2307/2007884
26. KpqC: Korean post-quantum crytography (2020). https://kpqc.or.kr/
27. Lyubashevsky, V.: Fiat-shamir with aborts: applications to lattice and factoring-based signatures. In: Advances in Cryptology - ASIACRYPT 2009. Lecture Notes in Computer Science, vol. 5912, pp. 598–616. Springer (2009). https://doi.org/10.1007/978-3-642-10366-7_35
28. Lyubashevsky, V., Ducas, L., Kiltz, E., Lepoint, T., Schwabe, P., Seiler, G., Stehle, D., Bai, S.: Crystal–dilithum. https://pq-crystals.org/dilithium/index.shtml
29. Miller, V.S.: Use of elliptic curves in cryptography. In: Advances in Cryptology – CRYPTO '85. Lecture Notes in Computer Science, vol. 218, pp. 417–426. Springer (1986). https://doi.org/10.1007/3-540-39799-X_31
30. Min, S., Yamamoto, G., Kim, K.: Weak property of malleability in NTRUSign. In: Proc. of ACISP 2004. LNCS, vol. 3108, pp. 379–390 (2004)
31. National Institute of Standards and Technology: FIPS 197: Advanced Encryption Standard (AES) (November 2001). https://nvlpubs.nist.gov/nistpubs/FIPS/NIST.FIPS.197.pdf
32. National Institute of Standards and Technology: FIPS 186-4: Digital Signature Standard (DSS) (July 2013). https://nvlpubs.nist.gov/nistpubs/FIPS/NIST.FIPS.186-4.pdf
33. National Institute of Standards and Technology: FIPS 203: Module-Lattice-Based Key-Encapsulation Mechanism Standard (August 2024). https://doi.org/10.6028/NIST.FIPS.203
34. National Institute of Standards and Technology: FIPS 204: Module-Lattice-Based Digital Signature Standard (August 2024). https://doi.org/10.6028/NIST.FIPS.204
35. National Institute of Standards and Technology: FIPS 205: Stateless Hash-Based Digital Signature Standard (August 2024). https://doi.org/10.6028/NIST.FIPS.205

References

36. Nguyen, P.Q., Regev, O.: Learning a parallelepiped: cryptanalysis of GGH and NTRU signatures. J. Cryptol. **22**(2), 139–160 (2009)
37. NIST: Post-quantum crytography (2016). https://csrc.nist.gov/projects/post-quantum-cryptography
38. Pornin, T., Prest, T.: More efficient algorithms for the NTRU key generation using the field norm. In: Lin, D., Sako, K. (eds.) Public-Key Cryptography – PKC 2019, pp. 504–533. Springer International Publishing, Cham (2019)
39. Prest, T.: Gaussian Sampling in lattice-based cryptography. Ph.D. thesis, École Normale Supérieure, Paris (2015)
40. Rivest, R.L., Shamir, A., Adleman, L.: A method for obtaining digital signatures and public-key cryptosystems. Commun. ACM **21**(2), 120–126 (1978). https://doi.org/10.1145/359340.359342
41. Schwabe, P., Avanzi, R., Bos, J., Ducas, L., Kiltz, E., Lepoint, T., Lyubashevsky, V., Schanck, J.M., Seiler, G., Stehle, D., Ding, J.: Crystal–kyber. https://pq-crystals.org/kyber/index.shtml
42. Shannon, C.E.: Communication theory of secrecy systems. Bell Syst. Tech. J. **28**(4), 656–715 (1949)
43. Shor, P.W.: Polynomial-time algorithms for prime factorization and discrete logarithms on a quantum computer. SIAM Rev. **41**(2), 303–332 (1999)
44. Stehlé, D., Steinfeld, R.: Making NTRU as secure as worst-case problems over ideal lattices. In: Paterson, K.G. (ed.) EUROCRYPT 2011. LNCS, vol. 6632, pp. 27–47. Springer, Heidelberg (2011)
45. Wikipedia: Harvest now, decrypt later (2023). https://en.wikipedia.org/wiki/Harvest_now_decrypt_later

Index

Symbols
ℓ_2-norm, 5
ℓ_∞-norm, 5
q-ary lattice, 7
ANTRAG, 19
DILITHIUM, 3
FALCON-512, 42
FALCON-1024, 42
KYBER, 3
MITAKA, 19
NTRUENCRYPT, 11
NTRUSIGN, 11
NtruSolve, 22
SOLMAE-512, 61
SOLMAE-1024, 61
SPHINCS+, 3

B
Bézout-like equation, 22

C
Closest Vector Problem (CVP), 11
Cyclotomic polynomials, 19
Cyclotomic ring, 19

D
Diffie and Hellman (DH), 2
Digital Signature Algorithm (DSA), 2
Digital Signatures (DS), 2
Discrete Fourier Transform (DFT), 8
Discrete Gaussians, 8

E
ElGamal, 2
Euler's totient function, 5

F
Fast Fourier Transform (FFT), 22
ffSampling, 14
Fiat-Shamir-with-aborts, 2
Full-rank matrix, 7

G
GGH, 11
Gram-Schmidt orthogonalization (GSO), 8

H
Hash-and-sign, 2

I
Inner product, 6

K
Key Encapsulation Mechanisms (KEM), 2

L
Lattices, 7
LDL* decomposition, 9

M
Matrices, 5

N
NTRU Lattices, 7
NTRU decision problem, 8
NTRU search problem, 8
Number fields, 5
Number Theoretic Transform (NTT), 17, 24
Number Theory aRe Us/N-th degree TRUncated polynomial (NTRU), 11

P
Post Quantum Cryptography (PQC), 3
Python, 27

Q
Quotient rings, 5

R
Ring lattices, 7
Rivest Shamir Adleman (RSA), 2

S
Scalars, 5
Shor, P.W., 2
Side-channel attacks, 79

V
Vectors, 5
Volume, of a lattice, 7